Conceptual and Technical Challenges for Quantum Gravity 2014 – Parallel Session: Noncommutative Geometry and Quantum Gravity

Journal of Physics: Conference Series Volume 634

Rome, Italy
8-12 September 2014

Editors:

Pierre Martinetti
Jean-Christophe Wallet
Giovani Amelino-Camelia

ISBN: 978-1-5108-0974-1
ISSN: 1742-6588

Printed from e-media with permission by:

Curran Associates, Inc.
57 Morehouse Lane
Red Hook, NY 12571

Some format issues inherent in the e-media version may also appear in this print version.

Printed by Curran Associates, Inc. (2015)

For permission requests, please contact the Institute of Physics
at the address below.

Institute of Physics
Dirac House, Temple Back
Bristol BS1 6BE UK

Phone: 44 1 17 929 7481
Fax: 44 1 17 920 0979

techtracking@iop.org

Additional copies of this publication are available from:

Curran Associates, Inc.
57 Morehouse Lane
Red Hook, NY 12571 USA
Phone: 845-758-0400
Fax: 845-758-2634
Email: curran@proceedings.com
Web: www.proceedings.com

Table of contents

Volume 634
Conceptual and Technical Challenges for Quantum Gravity 2014 – Parallel session: Noncommutative Geometry and Quantum Gravity
8–12 September 2014, Roma, Italy

Accepted papers received: 29 July 2015
Published online: 3 August 2015

Preface

Papers

Journal of Physics: Conference Series **634** (2015) 011001 doi:10.1088/1742-6596/634/1/011001

Quantum Gravity and Noncommutative Geometry

Editors: Pierre Martinetti[1], Jean-Christophe Wallet[2] (for the parallel session);
Giovani Amelino-Camelia[3] (for the conference).

Preface

The conference **Conceptual and Technical Challenges for Quantum Gravity** at *Sapienza University of Rome*, from 8 to 12 September 2014, has provided a beautiful opportunity for an encounter between different approaches and different perspectives on the quantum-gravity problem. It contributed to a higher level of shared knowledge among the quantum-gravity communities pursuing each specific research program.

There were plenary talks on many different approaches, including in particular string theory, loop quantum gravity, spacetime noncommutativity, causal dynamical triangulations, asymptotic safety and causal sets. Contributions from the perspective of philosophy of science were also welcomed.

In addition several parallel sessions were organized. The present volume collects contributions from the *Noncommutative Geometry and Quantum Gravity* parallel session [4], with additional invited contributions from specialists in the field.

Noncommutative geometry in its many incarnations appears at the crossroad of many researches in theoretical and mathematical physics:

- from models of quantum space-time (with or without breaking of Lorentz symmetry) to loop gravity and string theory,
- from early considerations on UV-divergencies in quantum field theory to recent models of gauge theories on noncommutative spacetime,
- from Connes description of the standard model of elementary particles to recent Pati-Salam like extensions.

This volume provides an overview of these various topics, interesting for the specialist as well as accessible to the newcomer.

[1]Università di Trieste, pmartinetti@units.it
[2]Université Paris-Sud 11 & CNRS, Jean-Christophe.Wallet@th.u-psud.fr
[3]Università di Roma Sapienza & INFN, amelino@roma1.infn.it
[4]partially funded by CNRS PEPS /PTI "Metric aspect of noncommutative geometry: from Monge to Higgs"

Peer review statement

All papers published in this volume of *Journal of Physics: Conference Series* have been peer reviewed through processes administered by the proceedings Editors. Reviews were conducted by expert referees to the professional and scientific standards expected of a proceedings journal published by IOP Publishing.

Beyond the Standard Model with noncommutative geometry, strolling towards quantum gravity

Pierre Martinetti

Università di Trieste, dipartimento di Matematica e Informatica, via Valerio 12/1, I-34127

E-mail: pmartinetti@units.it

Abstract. Noncommutative geometry in its many incarnations appears at the crossroad of many researches in theoretical and mathematical physics: from models of quantum spacetime (with or without breaking of Lorentz symmetry) to loop gravity and string theory, from early considerations on UV-divergencies in quantum field theory to recent models of gauge theories on noncommutative spacetime, from Connes description of the standard model of elementary particles to recent Pati-Salam like extensions. We list several of these applications, emphasizing also the original point of view brought by noncommutative geometry on the nature of time. This text serves as an introduction to the volume of proceedings of the parallel session "Noncommutative geometry and quantum gravity", as a part of the conference "Conceptual and technical challenges in quantum gravity" organized at the University of Rome *La Sapienza* in September 2014.

1. Introduction

Starting with early considerations of Bronstein, Mead, Wheeler, Pauli, Heisenberg and others (see [1] for a nice historical view on the subject), there exists a vast literature dealing with the reasons why putting together general relativity with quantum mechanics asks for a revision of the classical concept of spacetime. A common feature is the emergence of noncommutativity in the description of spacetime itself, as opposed to quantum mechanics alone where noncommutativity lives on the phase space. The appearance of spacetime noncommutativity has a wide range of motivations, from pure mathematics to phenomenological arguments. We shall not try to be exhaustive here, and simply point out some directions of research that have been through interesting developments in the last years. We shall make the distinction between *noncommutative spacetimes* intended as spaces whose coordinates no longer commute, and *spectral geometries* intended as a space whose algebra of functions is non-necessarily commutative.

Noncommutative spacetimes, as recalled in section 2, are obtained as a deformation of a usual space by trading the (commutative) coordinate functions x^μ, x^ν of a manifold with coordinate operators q^μ, q^ν satisfying non-trivial commutation relations. Besides the seminal quantum spacetime model of Doplicher, Fredenhagen, Roberts [2] treated in L. Tomassini talk, such spaces are present in many - if not all - approaches to quantum gravity, including loop quantum gravity, string theory (see P. Aschieri and R. Szabo text), as well as in more phenomenology-oriented models like doubly special relativity, as illustrated by S. Bianco in his presentation.

Journal of Physics: Conference Series **634** (2015) 012001 doi:10.1088/1742-6596/634/1/012001

Noncommutative spacetimes also emerged very early as a possible solution to ultraviolet divergencies in quantum field theory, especially in the work of Snyder [3]. Quantum field and gauge theories on noncommutative spacetimes have thus developed as a theory on their own, independently of any consideration on quantum gravity. Recent advances are presented in the texts of T. Juric, S. Meljanac, A, Samasarov one the one side, and J.-C. Wallet & A. Géré for noncommutative gauge theories [4] on the other.

Nevertheless, revisiting the work of Snyder in the light of nowadays quantum-gravity problematic offers an intriguing point of view, which is presented in V. Astuti paper.

Spectral geometries [5] is the subject of section 3. It consists in a generalization of Gelfand duality between locally compact spaces and C^*-algebras, so that to encompass all the aspects of Riemannian geometry [6] beyond topology. It furnishes a geometrical interpretation of the Lagrangian of the standard model of elementary particles [7, 8], as well as some possibilities to go beyond [9, 10]. Recent progress on that matter are reported in the contribution of A. Devastato. Finally F. Besnard. discusses the extension of this approach to the pseudo-Riemannian case.

2. Noncommutative spacetimes

2.1. Poincaré covariant spacetime

Spacetime as a pseudo-Riemannian manifold loses sense at Planck scale

$$\lambda_P = \sqrt{\frac{G\hbar}{c^3}} \simeq 1.6 \times 10^{-33} cm.$$

This is because an arbitrary accurate localization process requires to concentrate an arbitrary amount of energy in a small volume, yielding the creation of a black hole. To maintain an operational meaning to the measurement process, one should impose some non-zero minimal uncertainties in the simultaneous measurement of spacetime coordinates. This can be realized by promoting the coordinates functions x^μ to operators q^μ satisfying the non-commutative relation

$$[q_\mu, q_\nu] = Q_{\mu\nu}. \tag{1}$$

The behavior of (1) under a Poincaré transformation marks the difference between two classes of models, that both have given birth to an extended literature and many discussions, sometimes quite vivid. For simplicity, let us assume that the commutator of two coordinates is a central element (although some models of non-central commutators have been investigated, see in Tomassini's paper), that is

$$[q_\mu, q_\nu] = i\lambda_P^2 \theta_{\mu\nu} \mathbb{I} \tag{2}$$

where $\mathbb{I}$ is the identity operator in the Hilbert space on which the q^μ are represented and $\Theta = \{\theta_{\mu\nu}\}$ is an antisymmetric matrix. Obviously (2) is not invariant under the action of the Poincaré group

$$q_\mu \mapsto q'_\mu \doteq \Lambda_\mu^\alpha q_\alpha + a_\mu \mathbb{I} \qquad \Lambda \in SO(3,1), a \in \mathbb{R}^3 \tag{3}$$

since

$$[q'_\mu, q'_\nu] = [a_\mu, a_\nu]\mathbb{I} + q_\alpha([a_\mu \mathbb{I}, \Lambda_\nu^\alpha \mathbb{I}] - [a_\nu \mathbb{I}, \Lambda_\mu^\alpha \mathbb{I}]) + \Lambda_\mu^\alpha [q_\alpha, q_\beta] \Lambda_\nu^\beta \tag{4}$$

$$= i\lambda_P^2 \Lambda_\mu^\alpha \Lambda_\nu^\beta \theta_{\alpha\beta} \mathbb{I} \neq [q'_\mu, q'_\nu]. \tag{5}$$

One may ask that Θ transforms under the conjugate action of the Poincaré group, yielding the *Poincaré covariant* model of Doplicher, Fredenhagen, Roberts. The Planck length, viewed as the norm of the tensor $\lambda_P \Theta$ is Poincaré invariant, and there is no modification of the dispersion relation $E^2 = p^2c^2 + m^2c^4$. Applications of this model to cosmology are presented in **Luca Tomassini**'s *Noncommutative Friedmann spacetimes from Penrose-like inequalities.*

2.2. Deformed-Poincaré invariant spacetime

Alternatively one may require the *Poincaré invariance* of the relation (2) by imposing

$$[q'_\mu, q'_\nu] = i\lambda_P^2 \theta_{\mu\nu}. \tag{6}$$

This means that the symmetry group of the quantum space is no longer the Poincaré group, but a quantum group deformation of it (the so called θ-Poincaré quantum group), characterized by a non-trivial commutation relation for translations

$$[a_\mu, a_\nu] = i\theta_{\mu\nu} - i\theta^{\alpha\beta}\Lambda^\mu_\alpha \Lambda^\nu_\beta. \tag{7}$$

Under this symmetry, λ_P is again Lorentz invariant but there is now a modification of the dispersion relation

$$E^2 = p^2c^2 + m^2c^4 + f(m,p,E). \tag{8}$$

The possible experimental signature of (8) have been intensively explored, also for the Lie algebra- like noncommutativity (κ-Poincaré deformation)

$$[q_\mu, q_\nu] = \kappa\, \epsilon^\rho_{\mu\nu} q_\rho. \tag{9}$$

Such quantum-group deformations of Poincaré symmetries provide a useful tool to implement the idea of *Doubly Special Relativity*, that is the implementation of the Planck length λ_P as an invariant scale, in a similar manner as the speed of light c [11]. This has been interpreted later as a geometry where the space of momenta is curved (*Relative Locality*). Recent developments on that matter are treated in **Stefano Bianco** *Phenomenology from the DSR-deformed relativistic symmetries of 3D quantum gravity via the relative-locality framework.*

2.3. Quantum field theory on noncommutative space

Quantum field theories on noncommutative spacetime were put at the front of the scene by (open) string theory, when it was observed that D-brane world volume acquire a noncommutative deformation in the background of a non-zero B-field. This, together with new developments on closed string and nonassociative algebras, is recalled in **Paolo Aschieri & Richard J. Szabo** *Triproducts, nonassociative strar products and geometry of R-flux string compactifications.*

However one should not forget that quantum field and gauge theories on noncommutative spacetimes have been originally introduced independently of quantum gravity, as a tool to avoid ultraviolet divergencies. This was the original idea of Snyder, that has been somehow subsumed by renormalization. Nevertheless the idea that noncommutative spaces offer a beautiful ground to understand better quantum field and gauge theories, especially their renormalization properties, has been intensively investigated in the last decade. A scalar field model on a generalized κ-Minkowski space is presented in **Tajron Juric, Stjepan Meljanac, and Andjelo Samsarov** *Light-like κ-deformation and scalar field theory via Drinfeld twist.* The recent advances in gauge theory on noncommutative spacetimes [12] are reported in **Jean-Christophe Wallet** *Spectral theorem in noncommutative field theories: Jacobi dynamics.*

Finally, to go back to quantum gravity, let us also mention that Snyder's ideas re-thought from a quantum gravity perspective yields intriguing results. Recently, it has been used to question how much of the noncommutativity of the coordinates actually survive the description of physical systems. This is the object of **Valerio Astuti** *Covariant quantum mechanics applied to noncommutative geometry.*

3. Spectral geometry

On may question the physical meaning of the noncommutative coordinate q^μ in (1). As elements of the (abstract) polynomial algebra they generate, the spectrum of the q^μ's is not real, which makes their interpretation as physical observable problematic. A solution is to represent the q^μ's on some Hilbert space, as this is done in quantum mechanics. To do so, it is convenient to view the q_μ's as affiliated to the algebra of compact operators, as pointed out in the Doplicher-Fredenhagen-Roberts. But this indicates that the algebra generated by the coordinates may not be the most accurate tools to describe a quantum space, an algebra suitably associated to the q^μ's can do a better job. With this idea in minds, one has no reason to restrict one's attention to noncommutative deformations of commutative coordinates: there are much more noncommutative algebras to play with ! This idea is enforced by Gelfand duality between commutative C^*-algebras and locally compact spaces, which suggests that a natural definition of a *noncommutative geometry* is an object such that its "algebra of functions" (and not only its coordinates) is noncommutative.

3.1. The standard model and beyond

Connes' theory of spectral triples $(\mathcal{A}, \mathcal{H}, D)$ extends Gelfand duality, beyond topology, so that to encompass all the geometrical aspects of Riemannian geometry.

A spectral triple consists in a involutive algebra $\mathcal{A}$, a faithful representation on $\mathcal{H}$, a densely defined operator D on $\mathcal{H}$ such that $[D,a]$ is bounded and $a[D-\lambda\mathbb{I}]^{-1}$ is compact for any $a\in\mathcal{A}$ and $\lambda\notin$ Sp D. Imposing a set of further conditions, one defines *real* spectral triples, whose paradigmatic example is

$$(C^\infty(\mathcal{M}), L^2(\mathcal{M}, S), \partial\!\!\!/ = -i\gamma^\mu\partial_\mu), \tag{10}$$

that is the algebra of smooth functions on a closed spin manifold $\mathcal{M}$, acting on the Hilbert space of square integrable spinors, with $D = \partial\!\!\!/$ the Dirac operator. Conversely, one has the following reconstruction theorem [6]: given a real spectral triple $(\mathcal{A}, \mathcal{H}, D)$ with $\mathcal{A}$ unital commutative, then there exists a compact oriented Riemannian spin manifold $\mathcal{M}$ such that $\mathcal{A} = C^\infty(\mathcal{M})$. These conditions still makes sense for non-commutative $\mathcal{A}$, so that one defines a *noncommutative geometry* as a spectral triple $(\mathcal{A}, \mathcal{H}, D)$ where $\mathcal{A}$ is non necessarily commutative. To summarize:

$$\begin{array}{ccc} \text{commutative spectral triple} & \rightarrow & \text{noncommutative spectral triple} \\ \updownarrow & & \downarrow \\ \text{Riemannian geometry} & & \text{non-commutative geometry} \end{array}$$

Spectral triple turn out to be a powerful tool to describe the standard model of elementary particles from a purely geometric point of view. The starting point is to view spacetime no more as a manifold, as in general relativity, but as the product of a manifold by a matrix geometry. This allows to incorporate in the geometry the degrees of freedom of the standard model. More precisely, one considers the *almost-commutative* algebra

$$C^\infty(\mathcal{M}) \otimes \mathcal{A}_F, \tag{11}$$

where $\mathcal{A}_F$ is a finite dimensional algebra that carries the gauge group of the standard model (which is retrieved as the group of unitaries of $\mathcal{A}_F$). It acts on the space of fermions, that is $\mathbb{C}^{96}$ (two colored quarks, one neutrino and one electron make 8, that multiplies 2 for the chirality, another 2 for antiparticles and 3 for the number of generations). There is a finite dimensional Dirac operator D_F, that is a 96×96 matrix whose entries are the Yukawa coupling of the fermions and the mixing angles of quarks and neutrinos. A general formula for product of spectral triples yields the generalized Dirac operator $D := \partial\!\!\!/ \otimes \mathbb{I}_{96} + \gamma^5 \otimes D_F$. Bosonic fields are generated by the so-called fluctuations of the metric, that is the substitution of D with the covariant Dirac operator

$$D_A := D + A + JAJ^{-1} \tag{12}$$

where A is a selfadjoint element of the set of generalized 1-forms $\Omega^1_D := \{a^i[D, b_i]\}$. The asymptotic expansion $\Lambda \to \infty$ of the *spectral action* $\text{Tr}f(\frac{D}{\Lambda})$ [13] where Λ is a cutoff parameter and f a smooth approximation of the characteristic function of the interval $[0, 1]$ yields the bosonic Lagrangian of the standard model (including the Higgs) minimally coupled to Einstein-Hilbert action (in Euclidean signature).

In other terms the standard model appears as a purely gravitational theory, but on a (slightly) noncommutative space. As a bonus, the Higgs field comes out as the component of a connection in the noncommutative part of the geometry.

Practically, the spectral action provides relations between the parameters of the theory at a putative energy of unification. In particular the mass term of the Higgs appears as a function of the input of the models, namely the Yukawa couplings of fermions. Assuming the big-desert hypothesis, the running of this mass under the flow of the renormalization group yields a prediction for the mass of the Higgs of 170 GeV, a value ruled out by Tevatron in 2008.

Since then the Higgs-Brout-Englert boson has been discovered with a mass around 125 GeV. This mass is problematic, or at least intriguing, because it lies just below the threshold of stability, meaning that electroweak vacuum is a metastable state rather than a stable one. One solution to stabilize the electroweak vacuum is to postulate there exists another scalar field, called σ, suitably coupled to the Higgs. Chamseddine and Connes have noticed in [14] that taking into account this new scalar field in the spectral action, by promoting the Yukawa coupling of the right neutrino (which is one of the constant component of the matrix D_F) to a field,

$$k_R \to k_R\sigma, \tag{13}$$

then one obtains the correct coupling to the Higgs as well as a way to pull back the mass of the Higgs from 170 to 126GeV. In [10, 15], as reported in **Agostino Devastato** *Noncommutative geometry, grand symmetry and twisted spectral triple*, its is shown how the substitution (13) can be obtained as a fluctuation of the Dirac operator, but in a slightly modified version inspired by the notion of *twisted spectral triple* introduced previously by Connes and Moscovici. The field σ thus appears as a Higgs-like field associated to a spontaneous symmetry breaking to the standard model of a "grand symmetry" model where the spin degrees of freedom ($C^\infty(\mathcal{M})$ acting on the space of spinors) are mixed with the internal degrees of freedom ($\mathcal{A}_F$ acting on the space of particles).

3.2. Noncommutative space versus noncommutative space-time

Spectral triple provides a generalization of Riemannian geometry to the noncommutative setting, but there is no reconstruction theorem for pseudo-Riemannnian manifolds. Once computed the spectral action, one makes a Wick rotation $t \to it$, as done for instance in the path integral approach to quantum gravity. However one might like to make sense of Minkovskian signature from the beginning. In *Two roads to noncommutative causality*, **Fabien Besnard** presents a state of the art, including his own recent results, on various attempts to incorporate a causal structure in spectral triples.

Let us end this discussion by our own contribution, stressing the algebraic approach on how to put time into the game. This is the idea, sometimes advertised by Connes as "the heart of noncommutative geometry", that time involution is intrinsically contained within the noncommutativity of the algebra. Namely, given a von Neumann algebra $\mathcal{A}$ acting on an Hilbert space $\mathcal{H}$ together with a vector Ω in $\mathcal{H}$ cyclic and separating for the action of $\mathcal{A}$, one defines by

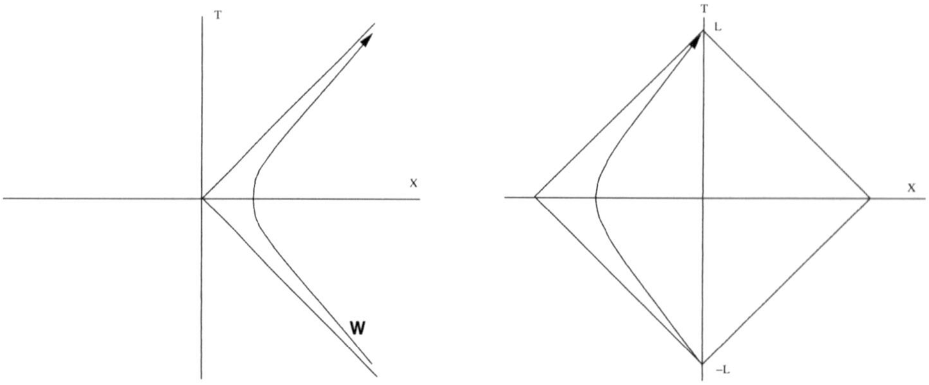

Figure 1. An orbit of the vacuum modular group for the algebra of local observables localized in the Rindler wedge (left) and in a double-cone of Minkowski spacetime (right).

Tomita-Takesaki modular theory a 1-parameter group of automorphism $\sigma_t^\Omega \in \text{Aut}(\mathcal{A})$. Connes has shown that the group obtained from another state Ω' differs from the former only by unitaries, that is

$$\sigma_t^{\Omega'} = U_t^{\Omega'\Omega}\sigma^\Omega(U_t^{\Omega'\Omega})^* \quad \forall t \in \mathbb{R} \tag{14}$$

where the unitary intertwining is given by Connes cocycle $U_t^{\Omega'\Omega}$. Hence there is a canonical group of outer automorphism σ_t canonically associated to the von Neumann algebra $\mathcal{A}$, where

$$\text{Out}(\mathcal{A}(\mathcal{O})) = \text{Aut}(\mathcal{A}(\mathcal{O}))/\text{Inn}(\mathcal{A}(\mathcal{O})).$$

The physical interpretation of this group as a time evolution is enforced by the fact that σ_Ω^t satisfies with respect to Ω the same properties as do in statistical physics the time evolution $e^{iHt} \cdot e^{-iHt}$ with respect to a Gibbs equilibrium state, namely the KMS condition.

An interesting framework to test the interpretation of the modular group as a real "physical" time is algebraic quantum field theory, where the von Neumann algebra is the algebra $\mathcal{A}(\mathcal{O})$ of local observables associated to an open region $\mathcal{O}$ of Minkowski spacetime. In particular, for $\mathcal{O}$ the Rindler Wedge, it is well know that the modular group of the vacuum vector is generated by the boosts. Hence it has a geometrical action whose orbits are the trajectories of uniformly accelerated observers. The KMS condition is interpreted as the vacuum being a thermal equilibrium state for this observer with a temperature proportional to the acceleration (see [16] for a recent critical view on this interpretation, though). A similar analysis holds for double-cone regions of Minkowski spacetime and yields a correction to the Unruh temperature for an observer with a finite life-time [17], see fig. 1.

Interestingly, a similar analysis for a double-cone in a bidimensional conformal field theory permits to compute explicitly the action of Connes cocycle. The field in the double-cone is determined by its components $\psi(x_1), \psi(x_2)$ on each interval on the boundary defining the double-cone (fig. 2). There is a state whose modular group has pure geometrical action $x_1(t), x_2(t)$ (but the orbit is not the trajectory of an observer with constant acceleration) while the modular group for the vacuum mixes this geometric action with a mixing of the components

$$\sqrt{\frac{dx_i}{d\zeta}}\,\sigma_t(\psi(x_i)) = \sum_{k=1,2} O_{ik}(t)\sqrt{\frac{dx_k}{d\zeta}}\,\psi(x_k(t))$$

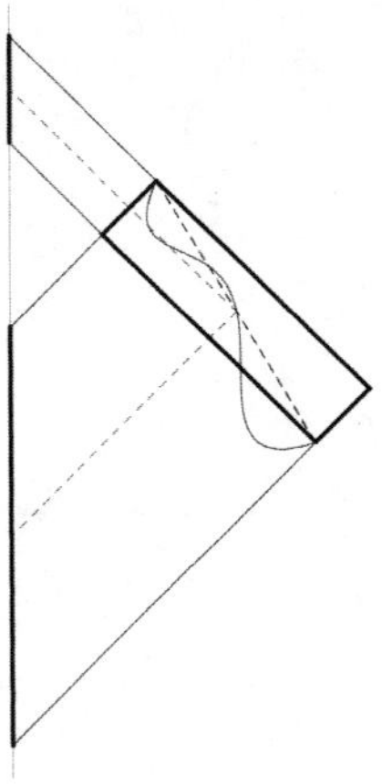

Figure 2. A modular orbit in $2D$-conformal theory with boundary.

where ζ is a suitable parametrization of the orbit. In this sense, the action of the unitary cocyle is non-geometric, and amounts to mix the components of the conformal field on the two intervals. For a further interpretation of this, see [18] and [20].

Besides quantum field theory, the hope is that this way of extracting a time flow from an algebra of observables and a state may be relevant in quantum gravity [19].

[1] Piacitelli G, Quantum Spacetime: a Disambiguation, 2010 *SIGMA* **6** 43 pages
[2] Doplicher S, Fredenhagen K and Robert J E, The quantum structure of spacetime at the Planck scale and quantum fields, 1995 *Commun.Math.Phys.* **172** 187–220
[3] Snyder H S, Quantized Space-Time, 1947 *Physical Review* **71**
[4] Wallet J-C, Noncommutative Induced Gauge Theories on Moyal Spaces, 2008 J. Phys.:Conf. Ser.**103** 012007, [*Preprint* arXiv:0708.2471]
[5] Connes A 1994 *Noncommutative Geometry* (Academic Press)
[6] Connes A, On the spectral characterization of manifolds, 2013 *J. Noncom. Geom.* **7** 1–82
[7] Connes A, Gravity coupled with matter and the foundations of Noncommutative Geometry, 1996 *Commun. Math. Phys.* **182** 155–176
[8] Chamseddine A H, Connes A and Marcolli M, Gravity and the standard model with neutrino mixing, 2007 *Adv. Theor. Math. Phys.* **11** 991–1089
[9] Chamseddine A H, Connes A and van Suijlekom W, Beyond the spectral standard model: emergence of Pati-Salam unification, 2013 *JHEP* **11** 132
[10] Devastato A, Lizzi F and Martinetti P, Grand Symmetry, Spectral Action and the Higgs mass, 2014 *JHEP* **01** 042
[11] Amelino-Camelia G, Doubly Special Relativity, 2002 *Nature* **418** 34–35
[12] Martinetti P, Vitale P and Wallet J-C, Noncommutative gauge theories on the Moyal plane as matrix models, 2013 *J. High Energy Phys.* **09** 051, [*Preprint* arXiv:1303.7185]
[13] Chamseddine A H and Connes A, The spectral action principle, 1996 *Commun. Math. Phys.* **186** 737–750
[14] Chamseddine A H and Connes A, Resilience of the spectral standard model, 2012 *JHEP* **09** 104
[15] Devastato A and Martinetti P, Twisted spectral triple for the standard model and spontaneous breaking of the grand symmetry, **arXiv 1411.1320**
[16] Buchholz D and Verch R, Macroscopic aspects of the Unruh effect. 2014 **arXiv 1412.5892**
[17] Martinetti P and Rovelli C, Diamond's temperature: Unruh effect for finite trajectories and the thermal time hypothesis, 2003 *Class. Quantum Grav.* **20** 4919–4931
[18] Rehren K H and Tedesco G, Multilocal fermionization, 2013 *Lett. Math. Phys.* **103** 19–36

[19] Connes A and Rovelli C, Von Neumann algebra automorphisms and time-thermodynamics relation in general covariant quantum theories, 1994 *Class. Quantum Grav.* **11**, 2899–2918.
[20] Martinetti P, Emergence of time in quantum gravity: is time necessarily flowing ? 2013, *Kronoscope* **13**, no. 1, 67–84.

Non commutative Friedmann spacetimes from Penrose-like inequalities

Luca Tomassini [1]
Università di Roma "Tor Vergata", Via della Ricerca Scientifica, 1, I-00133 Rome, Italy
E-mail: `tomassin@axp.mat.uniroma2.it`

Abstract. In this paper a generalisation of the Doplicher-Fredenhagen-Roberts (DFR) construction of a physically motivated non commutative Minkowski spacetime is generalised to the case of Friedmann flat expanding spacetimes. In particular, uncertainty relations quantitatively expressing limitations in the localisation of quantum observables are described and commutation relations implying them are derived from a general analysis. They reduce in a suitable sense to the DFR ones in the flat limit. Finally, it is shown that representations of these commutation relations do exist.

1. Introduction

The basic geometrical object of classical General Relativity is a four dimensional manifold (spacetime) M equipped with a lorentzian metric, all points of which can be labelled by (local) coordinates x^μ. Given this structure, relativistic quantum matter is introduced by means of quantum fields living on M. However, it has been since a long time recognised that such a picture becomes problematic at small scales, even though the very first motivation has been the need to get rid of the well known ultraviolet divergences (still) plaguing Quantum Field Theory. Indeed, in the very first speculations of Heisenberg, Sneyder and Yang of a possible non commutativity of spacetime there seems to have been no substantial direct physical motivation. It was only some years later, apparently with the work Mead [25], that the idea appeared that General Relativity itself could lead to limitations in the precision of spacetime position measurements. The idea was natural and compelling. To make observations in a given spacetime region we must use (say) radiation with comparable wavelength. In case the localisation region is very small (so that the energy density can be assumed very big) this should eventually lead to the formation of an event horizon, i.e. of a black hole ([3]). Thus, no information could come out of the region itself, making the measurement meaningless. Mead also gave the first quantitative estimate of such a lower bound, showing it to be of the scale of the Planck length, that is the utterly small quantity

$$\lambda_P = \sqrt{\frac{G\hbar}{c^3}} \simeq 1,6.10^{-33}\text{cm},$$

where c and G are respectively the speed of light and the gravitational constant. Still, though Mead's result was derived making use of a kind of Heisenberg's microscope argument, for many years to come none of his findings was interpreted as a signal of the non commutativity of

[1] Work supported by the **ERC Advanced Grant** 227458 OACFT *Operator Algebras & Conformal Field Theory.*

spacetime, nor any connection with the previously mentioned problems of Quantum Field Theory was explicitly indicated. Perhaps, among the reasons were on the one side the irrelevance of gravity in scattering experiments and on the other the lack of an explicit non commutative geometry as constructed thanks to the work [1] by Connes, Takesaki and many others.

Finally, the pieces were all taken together for the first time in the pioneering work of Doplicher, Fredenhagen and Roberts [15][2], who advocated the idea that the structure of non commutative spacetime Quantum Fields live on should be derived by (or at least not be in contrast with) established physical principles. They thus started from a clearcut formulation of Mead's result, eventually to take the form of a Principle of gravitational stability for localisation of events (PGSL):

- The gravitational field generated by the concentration of energy required to localise an event in spacetime should not be so strong to hide the event itself to any distant observer - distant compared to the Planck scale.

From it, they were able to derive the following bounds on the uncertainties of the coordinates of events for Minkowski spacetime:

$$c\Delta t\left(\Delta x^1 + \Delta x^2 + \Delta x^3\right) \geq \lambda_P^2, \tag{1}$$

$$\Delta x^1 \Delta x^2 + \Delta x^1 \Delta x^3 + \Delta x^2 \Delta x^3 \geq \lambda_P^2. \tag{2}$$

where the Δx^μ's, $\mu = 0, \ldots, 3$ should be interpreted as lengths of the edges of the relevant "localisation box". The quantities Δx^μ, $\mu = 0, \ldots, 3$ were then interpreted as the mean standard deviations

$$\Delta x^\mu \doteq \Delta_\omega x^\mu = \sqrt{\omega((x^\mu)^2) - \omega(x^\mu)^2} = \sqrt{(\phi, (x^\mu)^2\phi) - (\phi, x^\mu\phi)^2} \tag{3}$$

of hermitian operators x^μ, $\mu = 0, \ldots, 3$, acting on some Hilbert space $\tilde{\mathcal{H}}$ with scalar product $\widetilde{(\cdot,\cdot)}$ (here ω_ϕ is a so called vector state and $\phi \in \tilde{\mathcal{H}}$). These operators had to satisfy commutation relations such that (1) would follow, which were explicitly found. This is of course in complete analogy with Quantum Mechanics, but with the crucial difference that now the same mathematical objects are interpreted in a completely different way. The x^μ are to be regarded to as (global) non commutative coordinates on a non commutative Minkowski spacetime $\mathcal{E}$, which in turn was (as usual in non commutative geometry) identified with the C^*-algebra (see [6]) they generate (see below for more details).

To implement the PGSL, we need general conditions for the (non) formation of horizons on the relevant background. These (when they exist) are usually expressed in terms of some definition of the energy of the localised matter or radiation. In [15] the linear approximation of Einstein's equations was used to derive these conditions and the Heisenberg uncertainty relations where used to estimate the (local) energy. Later, several improvements followed in this respect which did not change the essential picture. First, in [30], where Penrose's Inequality was used to derive a more precise answer to 1) and accordingly stricter inequalities for the localisation uncertainties were derived. It is a remarkable fact that equations (1) could be recovered from them. This method has also the advantage to admit straightforward generalisations to general curved spacetimes. Accordingly, a modification of Penrose's inequality derived in [13, 14] for the case of Friedmann expanding spacetimes was used in [31] to derive uncertainty inequalities in the *flat* case. In both papers, however, the energy transfer to spacetime caused by the localisation process was still estimated in terms of the Heisenberg uncertainty relations. A completely different approach was taken in [18]. In this remarkable paper the ambiguities in the definition of energy were removed by making use of a relativistic scalar quantum field evaluated in a suitable state (and the backreaction fully taken into account using Raychaudhury equation).

[2] see also the interesting reviews [10, 12, 28].

The treatment, however, was limited to the case of spherically symmetric localisation regions and for this reason no uncertainty relations could be recovered. Still, all obtained results are in perfect quantitative agreement with (1).

One more important points have to be stressed concerning the road to Quantum Gravity proposed in [15] (see [19] for a further discussion). While being "semiclassical" in the sense that the existence of spacetime itself is taken as a starting assumption as opposed for example to Loop Quantum Gravity (see for example [8] for a comprehensive introduction), the DFR approach is fully quantum field theoretical in the sense that the above mentioned spacetime is considered as a way to describe the localisation properties of relativistic quantum fields. In other terms, while in non relativistic Quantum Mechanics "position" is position of (say) a single particle and is of course an observable, a relativistic field is nothing more (and less) than a way to attach physical degrees of freedom to a spacetime point. This means that the physical quantities the theory speaks about are precisely those degrees of freedom and spacetime points have to be seen as a mere labels or (if coordinates are introduced) parameters[3]. In the case of Quantum Field Theory on non commutative spacetime (*à la* DFR) these parameters are non commutative. Thus, the term "event" is used in the formulation of the PGSL above in the very same sense of Einstein: an event is *not* a physical entity but rather a place (specified by some parameters called coordinates) where a physical process could take place. Of course, quantum fields do not really make sense when evaluated at a point and it is usually assumed that all quantum observables localised in some spacetime region $\mathcal{O}$ may be constructed from the fields "smeared" with test functions f whose support is contained in $\mathcal{O}$, but this does not at all change the picture. Quantum fields on DFR non commutative Minkowski spacetime were actually constructed from the very beginning in [15] and their properties studied in a long series of papers (see for example [10, 11]).

This paper, in which we present ongoing work with G. Morsella [26], is organised as follows. In Section 2 we give a description of non commutative Minkowski spacetime. We include a brief discussion of quantum fields on it, with the aim to stress that these are the fundamental objects of the DFR approach to Quantum Gravity. In Section 3 we illustrate the use of Penrose-like inequalities to derive uncertainty relations. First we illustrate the method for the case of Minkowski spacetime, then use it for the case of general expanding flat Friedmann spacetimes. Sections 4 and 5 are dedicated to the derivation of commutation relation from which the Friedmann uncertainty relations follow. We will see on general grounds that in order to get the DFR model in the flat limit a new generator has to be added to the coordinates. Finally, in Section 6 we show that representations of our commutation relations indeed exist.

2. The Doplicher-Fredenhagen-Roberts model

2.1. DFR non commutative Minkowski spacetime

As it was mentioned in the introduction, for the case of Minkowski spacetime in [15] a C^*-algebra $\mathcal{E}$ was constructed whose (unbounded) selfadjoint generators x_μ have commutation relations such that the uncertainty inequalities (1) are satisfied in any state ω on $\mathcal{E}$ in their domain (*i.e.* such that $\omega(x_\mu^2) < \infty$). These commutation relations are

$$[x_\mu, x_\nu] = i\lambda_P^2 Q_{\mu\nu}, \qquad x^\mu = (x^\mu)^*, \tag{4}$$

and follow from an abstract analysis based on fairly general assumptions. Of these, the physically most significant is *covariance*, that is existence of an automorphic action of the (full) Poincaré group $\mathcal{P}$ on $\mathcal{E}$ such that

$$\alpha_{(\Lambda,a)}(x_\mu) = \Lambda_\mu^{\ \nu} x_\nu + a_\mu \mathbb{I}, \qquad \alpha_{(\Lambda,a)}(Q_{\mu\nu}) = \Lambda_\mu^{\ \mu'} \Lambda^{\nu'}_{\ \nu} Q_{\mu'\nu'}. \tag{5}$$

[3] Their status may be compared with the one of time in non-relativistic Quantum Mechanics.

This is to be contrasted with models of non commutative Minkowski spacetime in which actions of *quantum* versions of the Poincaré group are assumed and is justified by the expectation that global space motions should look the same in the large and in the small, while being the classical ones in the large[4]. Moreover, the so called "Quantum conditions" are required. They read

$$[x_\mu, Q_{\nu\rho}] = 0, \qquad Q^{\mu\nu}Q_{\mu\nu} = 0, \qquad (Q^{\mu\nu}(*Q)_{\mu\nu})^2 = 16\mathbb{I}. \tag{6}$$

where $\mathbb{I}$ is the identity operator, $(*Q)_{\mu\nu} = \epsilon_{\mu\nu\eta\rho}Q^{\eta\rho}$ and $\epsilon_{\mu\nu\eta\rho}$ is the totally antisymmetric tensor. The last two equality in terms of the electric and magnetic components $\boldsymbol{e} = (Q_{01}, Q_{02}, Q_{03})$ and $\boldsymbol{m} = (Q_{12}, Q_{32}, Q_{13})$ of Q read

$$\boldsymbol{e}^2 = \boldsymbol{m}^2 \qquad (\boldsymbol{e}\cdot\boldsymbol{m})^2 = \mathbb{I} \tag{7}$$

To understand what $\mathcal{E}$ and its generators look like, observe that because of their centrality (6) the operators $Q_{\mu\nu}$ have a joint spectrum Σ and because of covariance they reduce in an irreducible representation to an ordinary matrix

$$\sigma_{\mu\nu} = \mathrm{Ad}_{(\Lambda_\sigma)}(S) = (\Lambda^{\mu'}_{\mu}\Lambda^{\nu'}_{\nu}S_{\mu'\nu'}), \qquad S = (S_{\mu\nu}) = \begin{pmatrix} 0 & 0 & -1 & 0 \\ 0 & 0 & 0 & -1 \\ 1 & 0 & 0 & 0 \\ 0 & 1 & 0 & 0 \end{pmatrix}.$$

where $\mathrm{Ad}(\cdot)$ indicates the adjoint action of the Lorentz group $\mathscr{L}$. Correspondingly, we must have $x_\sigma = \Lambda_\sigma\tilde{x}$ with

$$\tilde{x}_0 = p_1,\ \tilde{x}_1 = p_2,\ \tilde{x}_2 = q_1,\ \tilde{x}_3 = q_2, \tag{8}$$

p_i, q_i being a pair of the usual Schrödinger operators. It is thus natural to define $\mathcal{E}$ through a generalised Weyl correspondence, namely as the (unique) C^*-completion of the Banach algebra of continuous functions from Σ to $L^1(\mathbb{R}^4, d^4\alpha)$ and vanishing at infinity equipped with the star-product, involution and norm

$$(f \star g)(\sigma, \alpha) = \int f(\sigma, \alpha')g(\sigma, \alpha - \alpha')e^{\frac{i}{2}\alpha_\mu\sigma^{\mu\nu}\alpha'_\nu}, \tag{9}$$

$$f^*(\sigma, \alpha) = \overline{f(\sigma, -\alpha)}, \qquad ||f|| = \sup_{\sigma\in\Sigma} ||f(\sigma, \cdot)||_1. \tag{10}$$

A fiberwise application of von Neumann's Uniqueness Theorem implies then the existence of a unique (up to unitary equivalence) representation of $\mathcal{E}$ acting on the direct integral (over $\mathscr{L}$) Hilbert space $\mathcal{H} = \int^\oplus d\Lambda\,(L^2(\mathbb{R})\otimes L^2(\mathbb{R})) \simeq \int^\oplus d\Lambda\, L^2(\mathbb{R}^2)$. This allows to prove that $\mathcal{E}$ is isomorphic to $\mathscr{C}_0(\Sigma, \mathcal{K})$, $\mathcal{K}$ being the compact operators.

According to the principles of non-commutative geometry (pure) states ω on $\mathcal{E}$ should be viewed as "non-commutative points" (see [23, 24] for a discussion on that matter in the DFR context). Although for general ω positivity and the commutation relations prevent pointlike localisation, the interesting class of *maximally localised* states minimising the quantity $\sum_\mu(\Delta(x_\mu))^2$ was introduced in [15]. Essentially, they are ground states of the Hamiltonian $H = 1/2(p_1^2 + q_1^2 + p_2^2 + q_2^2)$ of the two-dimensional harmonic oscillator, *i.e.* translated gaussian states (sometimes called coherent states). A maximally localised state ω is completely characterised by the averages $\omega(x_\mu) = \overline{x}_\mu$, $\overline{x} \in \mathbb{R}^4$. Moreover, for any $\overline{x} \in \mathbb{R}^4$ and $\mu = 0, \ldots, 3$ we have $\Delta_{\omega_{\overline{x}}}(x_\mu) = \lambda_P$: we see that, so to speak, non commutativity makes points thick. The maximally localised states are looked at as the only ones to have geometrical meaning. There are some more reasons to consider maximally localised states. First, we have for them a well defined classical limit for $\lambda_P \to 0$ which in [15] was identified as $M \times \Sigma_0$, with $\Sigma_0 \subset \Sigma$ a certain six-dimensional compact manifold. Second, they naturally arise in the construction of Wick products of quantum fields on our non commutative spacetime $\mathcal{E}$.

[4] But see [29] for a proof that these are always suitable restrictions of the DFR model.

2.2. Quantum Fields

The generalisation of quantum fields to the non commutative setting has attracted much attention but once again we will stick to the approach outlined in [15] for Minkowski spacetime.

By analogy with the definition of an ordinary field on commutative Minkowski spacetime, one considers

$$\phi(x) = \int_{\mathbb{R}^4} dk\, e^{ikx} \otimes \hat{\phi}(k) \tag{11}$$

where $\hat{\phi}(k)$ is just the Fourier transform of the field on commutative spacetime but now e^{ikx} are the generalised Weyl operators associated to the DFR non commutative coordinates. Thus ϕ is now to be interpreted as a map from states on the C^*-algebra of non commutative Minkowski spacetime $\mathscr{E}$ to smeared field operators. More explicitly

$$\omega \to \phi(\omega) = \langle \omega \otimes \mathbb{I}, \phi(x) \rangle = \int_{\mathbb{R}^4} dx\, \phi(x) \psi_\omega(x),$$

where the r.h.s. is a quantum field on the ordinary spacetime, smeared with the test function ψ_ω defined by $\hat{\psi}_\omega(k) = \langle \omega, e^{ikx} \rangle$. In other terms, (smeared) non commutative quantum fields are functions from a quantum spacetime[5] to a C^*-algebra $\mathscr{F}$ (the C^*-algebra generated by ordinary fields) and are described by elements affiliated to $\mathscr{E} \otimes \mathscr{F}$.

We conclude this subsection with two important facts. First, one has

$$[\phi(\omega), \phi(\omega')] = i \int d^4x d^4y\, \Delta(x-y) \psi_\omega(x) \psi_{\omega'}(y),$$

so that knowledge of the classical commutator entails knowledge of its non commutative spacetime counterpart one. This will be the later taken as the starting point for our (proposed) generalisation to non commutative Friedmann spacetimes. Second, the modification of test functions illustrated above will in general entail that the evaluation of products of fields in a state generally involve non-local expressions.

3. Uncertainty relations from Penrose-like inequalities

3.1. Minkowski spacetime

The main idea in [30] to overcome the obvious contradictions involved in the application of the linear approximation of General Relativity to derive no-horizon conditions and thus describe very strong gravitational fields was to make use of Penrose inequality (see [22] for a beautiful introduction to the subject and a comprehensive bibliography). Here, we formulate it in somewhat loose terms.

For asymptotically flat data, horizons form if and only if

$$A < 16\pi \frac{G^2}{c^4} M_{\mathrm{ADM}}^2, \tag{12}$$

where A is the proper area enclosing the collapsing matter of total ADM mass M_{ADM}.

As it is well known the ADM mass is nothing else the total mass, *i.e.* the proper (local) mass M_p together with the (negative) contribution due to the binding gravitational energy. However, since to obtain our uncertainty relations we use the Heisenberg uncertainty relations

$$\Delta x_j \Delta p_j \geq \frac{\hbar}{2},\ i,j = 1,2,3 \qquad \Delta t \Delta E \geq \frac{\hbar}{2}, \tag{13}$$

[5] Recall that "non commutative points" are states on a C^*-algebra.

we need a formulation of Penrose inequality in terms of a proper local quantity. Could we limit ourselves to the case of spherically symmetric collapsing matter, we would have at our disposal the theorems in [13, 14, 21], where the condition that horizons form is expressed only in terms of proper lengths and masses. This is not the case but luckily we only need a sufficient no horizon condition, as provided by inverting (12) and observing that that $M_{\text{ADM}} \leq M_p$ (from now on we write $M = M_p$). Thus we get:

$$A \geq 16\pi \frac{G^2}{c^4} M^2. \tag{14}$$

First of all, we need to evaluate the proper area A of the localising region. As a working approximation and following [30], we take the one of the background (that is the chosen spacetime *without* the experiment). If we choose cartesian coordinates and indicate by Δx_i, $i = 1, 2, 3$, the sides length of a parallelepiped, in view of the fact that our background is for now by hypothesis minkowskian, we may write

$$A = \frac{\pi}{3}(\Delta x^1 \Delta x^2 + \Delta x^1 \Delta x^3 + \Delta x^2 \Delta x^3), \tag{15}$$

with the numerical constant chosen so that with $\Delta x^1 = \Delta x^2 = \Delta x^3 = 2\Delta R$ we have $A = 4\pi\Delta R^2$. Recalling now that for a single particle we have $E^2 \geq c^2(p_1^2 + p_2^2 + p_3^2)$, we identify momenta with the their quantum counterparts Δp_i, $i = 1, 2, 3$ in (13) and substitute in (14). We get:

$$\Delta A \geq 12\lambda_P^4 \left[\left(\frac{1}{\Delta x^1}\right)^2 + \left(\frac{1}{\Delta x^2}\right)^2 + \left(\frac{1}{\Delta x^3}\right)^2 \right]. \tag{16}$$

where we write from now on $\Delta A = \Delta x^1 \Delta x^2 + \Delta x^1 \Delta x^3 + \Delta x^2 \Delta x^3$. After some manipulation (see [30]) we get the space-space uncertainty relation

$$(\Delta_\omega x^1)^2 (\Delta_\omega x^2)^2 (\Delta_\omega x^3)^2 \geq 12\lambda_P^4 \left(\Delta_\omega x^1 \Delta_\omega x^2 + \Delta_\omega x^1 \Delta_\omega x^3 + \Delta_\omega x^2 \Delta_\omega x^3\right). \tag{17}$$

whereas using the time-energy Heisenberg relation we obtain

$$(c\Delta_\omega t)^2 \left(\Delta_\omega x^1 \Delta_\omega x^2 + \Delta_\omega x^1 \Delta_\omega x^3 + \Delta_\omega x^2 \Delta_\omega x^3\right) \geq 12\lambda_P^4. \tag{18}$$

Here, we added the subscript ω to emphasise the dependence on states now that the inequalities have a quantum interpretation. For later reference, we mention that

$$\Delta_\omega x^1 \Delta_\omega x^2 \Delta_\omega x^3 \geq \sqrt[4]{12^3}\lambda_P^3, \tag{19}$$

must then hold [30], so that a lower bound for the volume appears. From it, the existence of a mean maximal mass-energy density for such states can be deduced.

Significantly, the DFR uncertainty relations could be recovered as a weakening[6] of relations (17),(18) by making use of the following simple algebraic inequalities

$$\begin{aligned} (a + b + c)^2 &\geq ab + bc + ac, \\ (ab + bc + ac)^3 &\geq a^2 b^2 c^2. \end{aligned} \tag{20}$$

It is thus reasonable to regard the DFR commutation relations as a reasonable approximation to commutation relations implementing (17),(18). In Subsection 3.2, we will use the same approximation to build a DFR-like quantum Friedmann expanding spacetime.

[6] In particular, the ratio $\Delta A/\Delta V$ is substituted by the smaller quantity $(\Delta A)^{-1/2}$. Thus, localisation regions with zero volume are allowed.

Journal of Physics: Conference Series **634** (2015) 012002 doi:10.1088/1742-6596/634/1/012002

3.2. Uncertainty relations for Friedmann expanding spacetimes

Consider now any Friedmann flat expanding spacetimes M and fix *comoving coordinates* $(t, \mathbf{x})$. Thus takes the form

$$ds^2 = dt^2 - a(t)^2(dx_1^2 + dx_2^2 + dx_3^2), \tag{21}$$

and the Hubble parameter is $H(t) = a'(t)/a(t)$. In a previous paper the following generalisation of Penrose isoperimetric inequality with was formulated:

- Black holes do not form if the (positive) excess of proper mass-energy δE inside a two-surface S of proper area $\Delta\mathscr{A}$ contained in a slice of constant universal time t_0 satisfies the inequality:

$$\sqrt{\Delta\mathscr{A}}\left(\frac{1}{4\sqrt{\pi}} + \frac{H\sqrt{\Delta\mathscr{A}}}{4\pi c}\right) \geq \frac{G}{c^4}\delta E. \tag{22}$$

Since we restrict to expanding cosmologies, we shall assume from now on $H \geq 0$. Notice that Penrose inequality for Minkowski spacetime is recovered when $H = 0$, so that the condition for no black hole formation is *stronger* in this case. We will regard at the second factor in the left hand side of (22) as a correction to the flat case brought about by the expansion.

If we consider a box-like localisation region with comoving edges $\Delta x_1, \Delta x_2, \Delta x_3$, we get

$$\Delta\mathscr{A} = a^2(t)(\Delta x_1\Delta x_2 + \Delta x_1\Delta x_3 + \Delta x_2\Delta x_3) = a^2(t)\Delta A.$$

Calculations completely analogous to those in [31] lead to the following inequalities:

$$a^2(t)\Delta A\left(\frac{1}{4\sqrt{3}} + \frac{H(t)\sqrt{a(t)^2\Delta A}}{12c}\right) \geq \frac{\lambda_P^2}{2}, \tag{23}$$

$$c\Delta t \cdot \sqrt{a^2(t)\Delta A}\min_{t\in\Delta t}\left\{\left(\frac{1}{4\sqrt{3}} + \frac{H(t)\sqrt{a(t)^2\Delta A}}{12c}\right)\right\} \geq \frac{\lambda_P^2}{2}. \tag{24}$$

We now solve it with respect to the comoving area ΔA. For the sake of clarity, we state the following

Lemma 3.1. Let $f = (x_0 - c/\sqrt{3}H\lambda_P)^2$, where

$$x_0 = \begin{cases} \frac{2c}{\sqrt{3}H\lambda_P}\cos\left[\frac{1}{3}\arccos\left(-1 + \frac{9\sqrt{3}H^2\lambda_P^2}{c^2}\right)\right], & \text{for } \Delta > 0, \\ -sign(q)\frac{2c}{\sqrt{3}H\lambda_P}\cosh\left[\frac{1}{3}arccosh|1 - \frac{9\sqrt{3}H^2\lambda_P^2}{c^2}|\right], & \text{for } \Delta < 0, \end{cases} \tag{25}$$

is the greatest solution of the cubic equation $x^3 + px + q = 0$ and

$$\Delta = \frac{c^4}{\sqrt{3}H^4\lambda_P^4}(\frac{9\sqrt{3}H^2\lambda_P^2}{c^2} - 2),$$

$$p = -\left(\frac{c}{H\lambda_P}\right)^2, \qquad q = -\frac{4\sqrt{3}c^3}{9H^3\lambda_P^3}\left(1 - \frac{9\sqrt{3}H^2\lambda_P^2}{c^2}\right).$$

Then, inequality (23) is equivalent to

$$\Delta A \geq \lambda_P^2\, a(t)^{-2} f\left(H(t)\right). \tag{26}$$

The function f is strictly positive and analytic as a function of the variable H. Moreover, $f \sim H^{-2/3}$ when $H \to +\infty$ and $f \to 2\sqrt{3}$ when $H \to 0$.

Journal of Physics: Conference Series **634** (2015) 012002 doi:10.1088/1742-6596/634/1/012002

Proof. An obvious application of the formulas for the solutions of cubic equations with respect to the unknown $\tilde{x} = \sqrt{a^2\Delta A/\lambda_P}$. The behaviour of the solution for $H \to +\infty$ and $H \to 0$ can also be recovered from equation (24). □

We observed that the same conditions are stronger for Minkowski spacetime. It is easily seen that they are actually the strongest possible, so that it is reasonable to expect that (suitable) intermediate cases model situations in which the effects of curvature are underestimated but still taken into account. We thus feel free to substitute from (23) into (24) to obtain

$$c\Delta t \cdot \sqrt{a^2\Delta A} \geq \max_{t\in\Delta t}\{a(t)\Delta A\} \geq \lambda_P^2 \max_{t\in\Delta t}\left\{a(t)^{-2}f\left(H(t)\right)\right\}. \tag{27}$$

This, together with (26), will be our starting point for quantisation. In fact, the time average over Δt of the positive functions a, f_1 being smaller than the corresponding maximum and using the first of inequalities (20) to evaluate $\sqrt{\Delta A}$, we consider the following uncertainty relations:

$$\Delta_\omega A \geq \frac{\lambda_P^2}{2}|\omega(a^{-2}f)|, \tag{28}$$

$$c\Delta_\omega t\left(\Delta_\omega x_1 + \Delta_\omega x_2 + \Delta_\omega x_3\right) \geq \frac{\lambda_P^2}{2}|\omega(a^{-1}f)|, \tag{29}$$

where again the subscript ω emphasises the we now regard the inequalities as holding for any state on our non commutative spacetime and for future convenience we rescaled Planck's constant λ_P.

Consider now, at the non commutative level, a coordinate transformation

$$(t, \mathbf{x}) \to (t' = w(t), \mathbf{x}' = \mathbf{x}), \tag{30}$$

where w is any diffeomorphism with $w' > 0$. Since we are working at planckian scales, the approximation

$$\Delta t(t') \simeq \frac{\Delta t}{\Delta t'}\Delta t' = \left(\frac{dt}{dt'}\right)\Delta t' \tag{31}$$

is more than reasonable and we can write

$$c\Delta t'\left(\Delta x_1 + \Delta x_2 + \Delta x_3\right) \geq \frac{\lambda_P^2}{2}\left(\frac{dt}{dt'}\right)^{-1}\left(a^{-1}f\right)\left(t(t')\right).$$

where the function f is the same as before. We thus obtain, with obvious notation,

$$\Delta_\omega A \geq \frac{\lambda_P^2}{2}|\omega(f)|, \tag{32}$$

$$c\Delta_\omega t'\left(\Delta_\omega x_1 + \Delta_\omega x_2 + \Delta_\omega x_3\right) \geq \frac{\lambda_P^2}{2}|\omega\left(\left(\frac{dt}{dt'}\right)^{-1}a^{-1}f\right)|. \tag{33}$$

In the specific case of conformal coordinates

$$(\tau, \mathbf{x}) = \left(\int_{t_0}^{t} a(s)^{-1}ds + t_0, \mathbf{x}\right),$$

this gives

$$\Delta_\omega A \geq \frac{\lambda_P^2}{2}|\omega(a^{-2}f)|, \tag{34}$$

$$c\Delta_\omega \tau\left(\Delta_\omega x_1 + \Delta_\omega x_2 + \Delta_\omega x_3\right) \geq \frac{\lambda_P^2}{2}|\omega(a^{-2}f)|. \tag{35}$$

Journal of Physics: Conference Series **634** (2015) 012002 doi:10.1088/1742-6596/634/1/012002

From now on, we shall work in conformal coordinates both because of the simple symmetry of the corresponding STUR (34) and (35) and in view of the forthcoming applications to quantum field theory.

In the next sections, we provide explicit realisations of (34) and (35) in terms of generators of a non commutative Friedmann C^*-algebra. Then, we will see that an implementation of the above time-diffeomorphism symmetry can be obtained in terms of unitary operators.

4. Quantum Friedmann expanding spacetime

Following the general ideas of the DFR approach to Quantum Field Theory on non commutative spacetimes, we regard the coordinates as noncommutative generators of a C^*-algebra $\mathcal{A}$ describing the relevant non commutative spacetime. We fix conformal coordinates $(x_0 = \tau, \mathbf{x}$. Our commutation relations will be

$$[x_\mu, x_\nu] = i\lambda_P^2 Q_{\mu\nu}, \qquad [Q_{\mu\nu}, \tau] = 0, \qquad [Q_{\mu\nu}, Q_{\lambda\rho}] = 0 \tag{36}$$

where the second and third equalities are motivated by the fact that, being Friedmann spacetimes spatially homogeneous, we do expect the commutators to depend only on the time coordinate. The covariance properties of our operators above will be specified below. For a sensible generalisation of the structures and properties of DFR Minkowski spacetime, it is natural to formulate the following additional requirements:

1) The STUR (34), (35) should be satisfied for a physically relevant class of states.
2) There should be a compatible continuous action of the isometry group $G = SO(3) \ltimes \mathbb{R}^3$ of Friedmann spacetimes.
3) We should in some suitable sense recover the DFR model in the flat limit $a \to 1$.
4) The commutative background spacetime $M = I \times \mathbb{R}^3$ should be recovered (perhaps as a factor) in the limit $\lambda_P \to 0$.

Of course, 1) and 2) will be fully satisfied. Concerning 3), one should not be too strict. As a matter of fact, not even in the commutative case one fully recovers Minkowski space from Friedmann spacetimes in the zero density, flat limit, and as far as we know there is no standard procedure circumvent this problem. One way to understand this is to think about the different isometry groups of these spacetimes, the one of Friedmann spacetime being of course smaller than Minkowski's (see the end of this section for a further discussion). We will not address 4) in this paper.

Remark 4.1. From now on we shall always assume we work in some representation π on some (separable) Hilbert space $\mathcal{H}$ in which the operators $\pi(x_0) = \pi(\tau), \pi(\mathbf{x}), \pi(Q_{\mu\nu})$ are selfadjoint and $[\pi(x_\mu), \pi(x_\nu)]$ have selfadjoint closures $i\lambda_P^2\pi(Q_{\mu\nu}(\tau))$[7]. Moreover, we assume the existence of a common invariant dense domain D for the operators $\pi(\mathbf{x}), \pi(Q_{\mu\nu})$ and $\alpha(\tau)$, where α is any complex valued smooth function on $\mathrm{sp}(\tau)$. In the following, we shall drop the symbol π unless necessary.

Requirement 2) above now becomes: the non commutative coordinates transform "classically" under the given unitary action of $SO(3) \ltimes \mathbb{R}^3$. That is, with $R \in SO(3)$ and $R_{\mu\nu}$ having the obvious meaning,

$$U(R)\mathbf{x}U(R)^{-1} = R\mathbf{x}, \quad U(R)Q_{\mu\nu}U(R)^{-1} = R_{\mu\rho}R_{\nu\sigma}Q^{\rho\sigma}.$$

[7] Throughout this operators will be said to commute when they strongly commute, that is have commuting spectral resolutions.

Moreover, for $\mathbf{a} \in \mathbb{R}^3$, we must have

$$U(\mathbf{a})\mathbf{x}U(\mathbf{a})^{-1} = \mathbf{x} + \mathbf{a}\mathbb{I}, \qquad U(\mathbf{a})Q_{\mu\nu}U(\mathbf{a})^{-1} = Q_{\mu\nu}.$$

The generator τ must of course be invariant under both classes of transformations. Then, for $Q(\tau)$ to be a 2-tensor under the action of $SO(3)$, it is necessary and sufficient that its electric and magnetic component $\boldsymbol{e} = (Q_{01}, Q_{02}, Q_{03})$ and $\boldsymbol{m} = (Q_{12}, Q_{32}, Q_{13})$ of Q transform as vectors.

As in the case of the DFR model, our non commutative Friedmann spacetimes will be (to a large extent) characterised by the invariant quantities which can be constructed out of the Q's. Thus we impose the following *Friedmann Quantum Conditions*:

$$\boldsymbol{e}^2 = h_1(\tau)^2\mathbb{I}, \quad \boldsymbol{m}^2 = h_2(\tau)^2\mathbb{I}, \quad \boldsymbol{e}\cdot\boldsymbol{m} = h_3(\tau)\mathbb{I}, \tag{37}$$

where the dot indicates the usual euclidean scalar product and the functions $h_i \colon \mathrm{sp}(\tau) \to \mathbb{R}$ are assumed to be smooth with $h_1, h_2 > 0$.

We will determine these function through the uncertainty relations (34), (35).

By Remark 4.1 the operators $h_1(\tau), h_2(\tau)$ have inverses $h_1(\tau)^{-1}, h_2(\tau)^{-1}$ leaving D invariant and by (36) they commute with the operators $Q_{\mu\nu}$. Hence we can consider the operators $\tilde{\boldsymbol{e}} = h_1(\tau)^{-1}\boldsymbol{e}, \tilde{\boldsymbol{m}} = h_2(\tau)^{-1}\boldsymbol{m}$. They leave D invariant and commute with each other by (36). Moreover, by the Friedmann Quantum Conditions they are *bounded*. An important role will be played by the joint spectrum Z of $\tilde{\boldsymbol{e}}, \tilde{\boldsymbol{m}}$. Writing $Z \ni \zeta = (e, m)$, we see that by covariance Z is a G-space with respect to the action of the group $SO(3)$ given by $\zeta = (e, m) \to \zeta' = (Re, Rm)$. Thus Z is included in the set of pairs of real three-vectors e, m satisfying

$$e^2 = 1, \quad m^2 = 1, \tag{38}$$

and hence, being closed, it is compact.

We now make one more assumption, which will have very important consequences.

- The operators $\tilde{\boldsymbol{e}}, \tilde{\boldsymbol{m}}$ are central, that is $[\tilde{\boldsymbol{e}}, x_i] = [\tilde{\boldsymbol{m}}, x_i] = 0$, with $i = 1, 2, 3$.

It follows that the commutation relation (36) may now be written, with obvious notation,

$$[\tau, \boldsymbol{x}] = i\lambda_P^2 h_1(\tau)\tilde{\boldsymbol{e}}, \qquad [\boldsymbol{x}, \boldsymbol{x}] = i\lambda_P^2 h_2(\tau)\tilde{\boldsymbol{m}}. \tag{39}$$

Now, notice that if we define the (strong) commutant of a densely defined operator X on $\mathcal{H}$ as

$$(X)'_s := \{B \in \mathcal{B}(\mathcal{H}) \,:\, BX \subset XB\},$$

we immediately see that the (strong) commutativity $\tilde{\boldsymbol{e}}, \tilde{\boldsymbol{m}}$ with the selfadjoint operators $\mathbf{x}$ implies that $\tilde{\boldsymbol{e}}, \tilde{\boldsymbol{m}} \in (x_i)'_s$, $i = 1, 2, 3$ and similarly replacing x_i with τ.

Proposition 4.2. Let the operators $\tau, x_i, \boldsymbol{e}, \boldsymbol{m}$, with $i = 1, 2, 3$, be as above. Then the uncertainty relations

$$c\Delta_\omega\tau \sum_{j=1}^{3} \Delta_\omega x_j \geq \frac{\lambda_P^2}{2}|\omega(h_1(\tau))|, \tag{40}$$

$$\sum_{1\leq j<k\leq 3} \Delta_\omega x_j \Delta_\omega x_k \geq \frac{\lambda_P^2}{2}|\omega(h_2(\tau))|, \tag{41}$$

hold for all vector states $\omega = \omega_\psi$, $\psi \in D$.

Journal of Physics: Conference Series **634** (2015) 012002 doi:10.1088/1742-6596/634/1/012002

Proof. Let $\mathcal{M}$ be the abelian von Neumann algebra generated by $\tilde{\boldsymbol{e}}, \tilde{\boldsymbol{m}}$. By [2, Thm. 2 of part II.6], one has a direct integral decomposition

$$\mathcal{H} = \int_Z^{\oplus} \mathcal{H}(\zeta) d\mu(\zeta),$$

such that $\mathcal{M}$ is the algebra of (bounded) diagonal operators. Since any one of $x_\mu = (\tau, x_i)$ is a closed operator, its strong commutant is weakly closed [7, Lemma 7.2.8], and therefore $\mathcal{M} \subset (x_\mu)'_s$. This implies, by [7, Prop. 12.1.7], that X also decomposes along the direct integral

$$x_\mu = \int_Z^{\oplus} X(\zeta) d\mu(\zeta), \tag{42}$$

with selfadjoint $x_\mu(\zeta)$ on $\mathcal{H}(\zeta)$ such that $D(x_\mu) = \{\psi \in \mathcal{H} : \psi(\zeta) \in D(x_\mu(\zeta)) \text{ for a.e. } \zeta \in Z\}$. Therefore, with $\omega_\zeta := \omega_{\psi(\zeta)}$, $\psi \in D$, an argument similar to that in [15, Prop. 3.3] shows that

$$\Delta_\omega \tau \Delta_\omega x_j \geq \int_Z \Delta_{\omega_\zeta} \tau(\zeta) \Delta_{\omega_\zeta} x_j(\zeta) d\mu(\zeta)$$

and similarly for $\Delta_\omega x_j \Delta_\omega x_k$. Moreover

$$\int_Z \sum_{k=1}^{3} \Delta_{\omega_\zeta}(\tau) \Delta_{\omega_\zeta}(x_i) d\mu(\zeta) \geq \int_Z \sqrt{\sum_{k=1}^{3} \Delta_{\omega_\zeta}(\tau)^2 \Delta_{\omega_\zeta}(x_i)^2 d\mu(\zeta)} \geq$$
$$\geq \frac{1}{2} \int_Z \|\omega_\zeta([\tau(\zeta), \boldsymbol{x}(\zeta)])\| d\mu(\zeta)$$

Similarly,

$$\int_Z \sum_{1\leq j<k\leq 3} \Delta_{\omega_\zeta}(x_i) \Delta_{\omega_\zeta}(x_k) d\mu(\zeta) \geq \int_Z \sqrt{\sum_{1\leq j<k\leq 3} \Delta_{\omega_\zeta}(x_i)^2 \Delta_{\omega_\zeta}(x_k)^2 d\mu(\zeta)} \geq$$
$$= \frac{1}{2} \int_Z \sqrt{\sum_{1\leq j<k\leq 3} \left|\omega_\zeta([x_j(\zeta), x_k(\zeta)])\right|^2 d\mu_\omega(\zeta)}.$$

It is also clear that $\mathcal{M}$ commutes with the spectral projections of τ, which easily implies $\mathcal{M} \subset (F(\tau))'_s$, i.e., the decomposability of $h_1(\tau)$ too. Then, from the obvious relation

$$\int_Z^{\oplus} [\tau(\zeta), x_i(\zeta)] \psi(\zeta)\, d\mu(\zeta) = [\tau, x_j]\psi = i h_1(\tau) e_i \psi$$
$$= \int_Z^{\oplus} i h_1(\tau)(\zeta) \tilde{e}_i(\zeta) \psi(\zeta)\, d\mu(\zeta),$$

valid for $\psi \in D$, there follows, for a.e. $\zeta \in Z$,

$$\|\omega_\zeta([\tau(\zeta), \boldsymbol{x}(\zeta)])\| = \|\omega_\zeta(h_1(\tau)(\zeta)) \tilde{\boldsymbol{e}}(\zeta)\| = |\omega_\zeta(h_1(\tau)(\zeta))|,$$

having used the fact that $\tilde{e}_i$ is diagonal and $\|\tilde{\boldsymbol{e}}(\zeta)\| = 1$. By a similar reasoning one also has

$$\sum_{1\leq j<k\leq 3} \left|\omega_\zeta([x_i(\zeta), x_j(\zeta)])\right|^2 = \|\omega_\zeta(h_2(\tau)(\zeta)) \tilde{\boldsymbol{m}}(\zeta)\|^2 = |\omega_\zeta(h_2(\tau)(\zeta))|^2.$$

Finally, we recall that by assumption h_1, h_2 are positive functions. □

Thanks to the previous result, we see that the uncertainty relations (34), (35) are satisfied if we set $h_1 = h_2 = a^{-2}f(H) \doteq F$ (notice that F is regular and stricly positive). We also observe that the result holds for *any* a, H.

The above assumptions on the commutators $Q_{\mu\nu}$ have interesting consequences, but to show this we will need the following technical result (see [26]).

Proposition 4.3. Let τ, x, E be selfadjoint operators on a Hilbert space $\mathcal{H}$, $I \subset \mathbb{R}$ an interval containing the spectrum of τ, $F : I \to \mathbb{R}$ a Borel function, and assume that E strongly commutes with τ, x and that there is a dense subspace $D \subset \mathcal{H}$ contained in $D([\tau, x]) \cap D(F(\tau)E)$ such that

$$[\tau, x]\psi = iF(\tau)E\psi, \qquad \psi \in D.$$

Assume moreover that $\alpha : I \to \mathbb{R}$ is a C^1 function and that the dense subspace D is invariant for $x, \tau, (\tau \pm i)^{-1}$ and is contained in $D(\alpha(\tau)) \cap D(\alpha F(\tau)E) \cap D(\alpha' F(\tau)E)$. Then there holds, for all $\psi \in D$,

$$[\alpha(\tau), x]\psi = i\alpha'(\tau)F(\tau)E\psi.$$

Now we are ready to prove the following

Proposition 4.4. Let the covariant generators $\tau, \mathbf{x}, \boldsymbol{e}, \boldsymbol{m}$ satisfy (39) and the assumptions of Remark 4.1. Then $\tilde{\boldsymbol{e}} \cdot \tilde{\boldsymbol{m}} = 0$.

Proof. From equations (39) and the assumed centrality of $\tilde{\mathbf{e}}, \tilde{\boldsymbol{m}}$, we see that Proposition 4.3 applies. Thus

$$[\boldsymbol{e}, x_i] = i\lambda_P^2 F'(\tau)F(\tau)\tilde{\mathbf{e}}\tilde{e}_i, \qquad [\boldsymbol{m}, x_i] = i\lambda_P^2 F'(\tau)F(\tau)\tilde{\boldsymbol{m}}\tilde{e}_i.$$

It follows that the only non trivial Jacobi identity for the x_μ is satisfied if

$$F(\tau)F'(\tau)\tilde{\mathbf{e}} \cdot \tilde{\boldsymbol{m}} = 0. \tag{43}$$

The operator on the left hand side being a joint functional calculus with a regular function, one concludes that such function must be identically vanishing on the joint spectrum of the operators. The result thus follows from $F, F' \neq 0$. □

In view of Requirement 3) above, we regard this as a *no go* result. In fact, taking into account that $F \to 1$ in the flat limit, we see that the DFR model cannot be recovered. In particular, while we expect that the first of the DFR Quantum Conditions (6) can be met, this is clearly impossible for the second. Instead, we see we recover a restriction to $SO(3)$ symmetry (we neglect translations) of the dilation covariant DFR-like model studied in [27], with a restriction to the corresponding Z (which now happens to be as in (58)) of the joint spectrum of the analogs of the operators $\boldsymbol{e}, \boldsymbol{m}$. Should we enlarge our symmetry group to the full Lorentz group we would be forced to do so with Z too, thus losing the minimal length λ_P. This problem will be addressed in the next section.

5. A quantum FRW with (almost) DFR Minkowskian limit

In order to comply with requirement 3) in Section 4, in the spirit of [16] we add a further auxiliary generator S to the (conformal) coordinates τ, x_i, $i = 1, 2, 3$, and assume commutation relations of the form

$$[\tau, x_i] = i\alpha(\tau)\tilde{e}_i, \qquad [x_i, x_j] = i\varepsilon_{ijk}(\beta(\tau)\tilde{m}_{1k} + \gamma(\tau)\tilde{m}_{2k}S) \tag{44}$$

$$[\tau, S] = 0, \qquad [x_i, S] = iC_i. \tag{45}$$

with $\tilde{e}_i, \tilde{m}_{ai}, C_i$, $i = 1, 2, 3$, $a = 1, 2$, central elements, and α, β, γ given regular functions. As in Section 4, we will assume that the assumption of Remark 4.1, obviously extended to the operators $\boldsymbol{C}, S$ are satisfied. In particular, the operators $\tilde{\boldsymbol{e}}, \tilde{\boldsymbol{m}}_1, \tilde{\boldsymbol{m}}_2$ are assumed to be central.

As before, the arbitrariness in the functions α, β, γ is restricted by the necessity of satisfying the Jacobi identities and Proposition 4.3 guarantees that the formal manipulations are indeed allowed. They imply

$$\beta'(\tau)\alpha(\tau)\tilde{\boldsymbol{m}}_1 \cdot \tilde{\boldsymbol{e}} - \gamma(\tau)\tilde{\boldsymbol{m}}_2 \cdot \boldsymbol{C} + \gamma'(\tau)\alpha(\tau)(\tilde{\boldsymbol{m}}_2 \cdot \tilde{\boldsymbol{e}})S = 0.$$

Once again (see the proof of Proposition 4.4) the operator on the left hand side are a joint functional calculus with a regular function, so that such function must be identically vanishing and therefore one gets the equations

$$\beta'(\tau)\alpha(\tau)\tilde{\boldsymbol{m}}_1 \cdot \tilde{\boldsymbol{e}} = \gamma(\tau)\tilde{\boldsymbol{m}}_2 \cdot \boldsymbol{C}, \qquad \gamma'(\tau)\alpha(\tau)(\tilde{\boldsymbol{m}}_2 \cdot \tilde{\boldsymbol{e}}) = 0.$$

Discarding the solution $\gamma =$ const of the second equation, this implies the conditions

$$\beta'(\tau)\alpha(\tau) = k\gamma(\tau), \qquad \tilde{\boldsymbol{m}}_2 \cdot \boldsymbol{C} = k\tilde{\boldsymbol{m}}_1 \cdot \tilde{\boldsymbol{e}}, \qquad \tilde{\boldsymbol{m}}_2 \cdot \tilde{\boldsymbol{e}} = 0, \tag{46}$$

for a suitable constant $k \in \mathbb{R}$. Without going further in the analysis of the consequences of (46), we consider the following particular solution

$$\begin{gathered} \alpha(\tau) = \beta(\tau) = F(\tau), \quad \gamma(\tau) = -F'(\tau)F(\tau), \\ \tilde{\boldsymbol{e}} = \tilde{\boldsymbol{m}}_1, \quad \tilde{\boldsymbol{m}} := \tilde{\boldsymbol{m}}_2 = -\boldsymbol{C}, \quad \tilde{\boldsymbol{e}}^2 = \tilde{\boldsymbol{m}}^2 = 1, \quad \tilde{\boldsymbol{e}} \cdot \tilde{\boldsymbol{m}} = 0, \end{gathered} \tag{47}$$

with F the right hand side of the STUR (26). This choice is motivated by a result completely analogous to Proposition 4.2, that is one can prove the inequalities

$$c\Delta_\omega \tau \sum_{j=1}^{3} \Delta_\omega x_j \geq \frac{\lambda_P^2}{2} |\omega(F(\tau))|, \tag{48}$$

$$\sum_{1 \leq j < k \leq 3} \Delta_\omega x_j \Delta_\omega x_k \geq \frac{\lambda_P^2}{2} \left(\big|\omega\big(F(\tau)\big)\big|^2 + \big|\omega\big(F(\tau)F'(\tau)S\big)\big|^2 \right) \geq \lambda_p^2 |\omega(F(\tau))|^2. \tag{49}$$

Moreover, it will be shown below that there are representations in which one can choose states with arbitrary localisation properties such that $\omega\big(F(\tau)F'(\tau)S\big) = 0$. We also point out the fact that on Minkowski spacetime F is a constant, and therefore (44) and (48), (49) reduce to the corresponding relations for the DFR model restricted to $Z^{(1)} = \{\zeta = (e, m) : e = m, e^2 = 1\} \cong \mathbb{S}^2$, a single orbit under the action of $SO(3)$. Again $(e, m) \mapsto (Re, Rm)$, $R \in SO(3)$.

All in all, the non zero commutation relations are thus, with obvious notation,

$$[\tau, \boldsymbol{x}] = iF(\tau)\tilde{\boldsymbol{e}}, \quad [\boldsymbol{x}, \boldsymbol{x}] = iF(\tau)(\tilde{\boldsymbol{e}} - F'(\tau)\tilde{\boldsymbol{m}}S) \tag{50}$$

$$[\tau, S] = 0, \quad [\boldsymbol{x}, S] = -i\tilde{\boldsymbol{m}}. \tag{51}$$

They can be rewritten in a particularly convenient form, as we now show. First, define the new central quantity $\tilde{\boldsymbol{n}} = \tilde{\boldsymbol{e}} \times \tilde{\boldsymbol{m}}$ (the symbol '$\times$' indicates here the exterior product). Then, an easy calculation shows that the scalar (with respect to the action of space rotations) quantities $\tilde{\boldsymbol{e}} \cdot \boldsymbol{x}, \tilde{\boldsymbol{m}} \cdot \boldsymbol{x}, \tilde{\boldsymbol{n}} \cdot \boldsymbol{x}$ satisfy

$$\begin{gathered} [\tau, \tilde{\boldsymbol{e}} \cdot \boldsymbol{x}] = iF(\tau), \quad [\tau, \tilde{\boldsymbol{m}} \cdot \boldsymbol{x}] = [\tau, \tilde{\boldsymbol{n}} \cdot \boldsymbol{x}] = [\tilde{\boldsymbol{e}} \cdot \boldsymbol{x}, \tilde{\boldsymbol{m}} \cdot \boldsymbol{x}] = 0, \\ [\tilde{\boldsymbol{e}} \cdot \boldsymbol{x}, \tilde{\boldsymbol{n}} \cdot \boldsymbol{x}] = iF(\tau)F'(\tau)S, \quad [\tilde{\boldsymbol{m}} \cdot \boldsymbol{x}, \tilde{\boldsymbol{n}} \cdot \boldsymbol{x}] = iF(\tau), \\ [\tilde{\boldsymbol{e}} \cdot \boldsymbol{x}, S] = [\tilde{\boldsymbol{n}} \cdot \boldsymbol{x}, S] = 0, \quad [\tilde{\boldsymbol{m}} \cdot \boldsymbol{x}, S] = -i\mathbb{I}, \\ [\tilde{\boldsymbol{e}} \cdot \boldsymbol{x}, F(\tau)S] = -iF(\tau)F'(\tau)S, \end{gathered}$$

from which we see that

$$[\tilde{\boldsymbol{e}}\cdot\boldsymbol{x},\tilde{\boldsymbol{n}}\cdot\boldsymbol{x}+F(\tau)S]=[\tilde{\boldsymbol{m}}\cdot\boldsymbol{x},\tilde{\boldsymbol{n}}\cdot\boldsymbol{x}+F(\tau)S]=[t,\tilde{\boldsymbol{n}}\cdot\boldsymbol{x}+F(\tau)S]=0. \tag{52}$$

Setting $y=\tilde{\boldsymbol{n}}\cdot\boldsymbol{x}+F(\tau)S$, we summarise our findings:

$$[\tau,\tilde{\boldsymbol{e}}\cdot\boldsymbol{x}]=iF(\tau),\quad [\tilde{\boldsymbol{m}}\cdot\boldsymbol{x},S]=-i\mathbb{I}, \tag{53}$$

all other commutators being zero. Equations (50), (51) thus decouple and this will be very useful when explicitly constructing concrete realisations. We now turn to discuss this problem.

6. Existence of representations

We will consider first relations (53) and then show how to get (50), (51). It should be clear that a simpler version of the construction below applies for Section 4.

The idea is that a solution is obtained by taking $\mathcal{H}=L^2(\mathbb{R}^2)$ (with the Lebesgue measure $d\xi_1 d\xi_2$) and defining

$$\tau:=\Phi(\xi_1),\quad \tilde{\boldsymbol{e}}\cdot\boldsymbol{x}:=i\frac{\partial}{\partial\xi_1}, \tag{54}$$

$$\tilde{\boldsymbol{m}}\cdot\boldsymbol{x}:=\xi_2,\quad y:=c\,\mathbb{I},\quad S:=i\frac{\partial}{\partial\xi_2}, \tag{55}$$

with $c\in\mathbb{R}$ and where a function of (ξ_1,ξ_2) is of course understood as the corresponding multiplication operator. The function Φ is given by the equations (see (53))

$$-(\xi-\underline{\xi})=\int_{\Phi(\underline{\xi})}^{\Phi(\xi)}\frac{ds}{F(s)}=\tilde{G}(\Phi(\xi))\;\rightarrow\;\Phi(\xi)=\tilde{G}^{-1}(-\xi), \tag{56}$$

because $\tilde{G}$ is invertible as a function of Φ (this follows from strict positivity of the function F). Finally, we point out that we will later obtain an invariant domain so that all the assumptions in Remark 4.1 from $D=C_c^\infty(\mathbb{R}^2)$. While irreducibility is obvious, the Remark below suggests that in general uniqueness of the realisation under the given assumptions is far from granted. However, we believe that this should be considered a problem only in case the different model have different *physical consequences* and we believe (but for now cannot prove) this is not the case.

Remark 6.1. As a matter of fact, the above construction of the operator τ does not work for arbitrary (increasing) expansion parameters $a(t(\tau))$ (see (21) and below). The point is that the function $\Phi(\xi)$ above as given by (56) might well diverge for *finite* values of ξ and thus be defined only in some open interval. The situation is best illustrated if we take, in the limit of small values of the universal time t, $a(t)\sim t^\gamma$ with $\gamma>1$. It follows that

$$\tau\sim\int_{t_0}^{t}a(s)^{-1}ds\sim\frac{1}{1-\gamma}\left(t^{1-\gamma}-t_0^{1-\gamma}\right),\qquad\Longrightarrow\qquad F(t(\tau))\sim\left((1-\gamma)\tau+t_0^{(1-\gamma)}\right)^{-\frac{2+3\gamma}{3(1-\gamma)}},$$

taking into account that $H\sim t^{-1}$ and $F\sim H^{-2/3}$ (see Lemma 3.1). A simple calculation then shows that

$$\Phi(\xi_1)=\frac{5(1-\gamma)}{3}\left(-\xi+\underline{\xi}+\frac{3(1-\gamma)}{5}\Phi(\underline{\xi})^{\frac{5}{3(1-\gamma)}}\right)^{\frac{3(1-\gamma)}{5}},$$

which makes the divergence at finite ξ apparent.

Fortunately, this problem may be circumvented as follows. In case the interval of definition of Φ is bounded, say $(\underline{\xi}_1, \underline{\xi}_2)$, then we simply extend the function Φ by periodicity. The Hilbert space will again be $L^2(\mathbb{R})$ and the derivative the usual Schrödinger operator, while the analog of D above will be given by smooth functions of compact support vanishing sufficiently fast with all their derivatives at the points $\underline{\xi}_1 \pm n(\underline{\xi}_2 - \underline{\xi}_1)$, with $n \in \mathbb{N}$. In case it is semibounded from above, say$(-\infty, \underline{\xi})$, we extend the function Φ to all of $\mathbb{R}$ by a reflection, that is we take $\tilde{\Phi} = -\Phi\left(-(\xi + 2\underline{\xi})\right)$ and redefine Φ in the obvious way. All goes as before, but D will be the smooth functions vanishing sufficiently fast with all their derivatives at the point $\underline{\xi}$. The case of an interval semibounded from below can be treated similarly.

So to say, this construction amounts to patching several copies of our Friedman universe one beside the other, a fact which should have no physical consequences as far as they do not "talk" to each other.

We now note that, given a realisation through operators $\tau, \boldsymbol{x}, S, \tilde{\boldsymbol{e}}, \tilde{\boldsymbol{m}}$, and an element $(\boldsymbol{a}, R) \in \mathbb{R}^3 \rtimes SO(3) =: G$ of the isometry group of Friedmann spacetime, a new realisation on the same Hilbert space is obtained by

$$\tau^{(\boldsymbol{a},R)} := \tau, \quad \boldsymbol{x}^{(\boldsymbol{a},R)} := R\boldsymbol{x} + \boldsymbol{a}\mathbb{I}. \quad S^{(\boldsymbol{a},R)} := S,$$
$$\tilde{\boldsymbol{e}}^{(\boldsymbol{a},R)} := R\tilde{\boldsymbol{e}}, \quad \tilde{\boldsymbol{m}}^{(\boldsymbol{a},R)} := R\tilde{\boldsymbol{m}}.$$

Therefore, we will say that a realisation is $\mathbb{R}^3 \rtimes SO(3)$-covariant if there is a unitary strongly continuous representation U of $\mathbb{R}^3 \rtimes SO(3)$ on $\mathcal{H}$ such that

$$U(\boldsymbol{a}, R)XU(\boldsymbol{a}, R)^* = X^{(\boldsymbol{a},R)},$$

for $X = \tau, \boldsymbol{x}, S, \tilde{\boldsymbol{e}}, \tilde{\boldsymbol{m}}$. Moreover, we call a realisation irreducible if any bounded operator $B \in \mathcal{B}(\mathcal{H})$ which commutes with $\tilde{\boldsymbol{e}}, \tilde{\boldsymbol{m}}$ and such that

$$B \in (S)'_s, \quad B \in (\tau)'_s, \quad B \in (\boldsymbol{x})'_s, \tag{57}$$

is a multiple of the identity. Therefore as expected in an irreducible realisation $\tilde{\boldsymbol{e}} = e\mathbb{I}$, $\tilde{\boldsymbol{m}} = m\mathbb{I}$ with (e, m) belonging to the manifold

$$Z^S := \{(e, m) \in \mathbb{R}^6 \,:\, e^2 = m^2 = 1,\, e \cdot m = 0\}. \tag{58}$$

Being Z^S a single orbit under the action $(e, m) \mapsto (Re, Rm)$, $R \in SO(3)$, we can assume that

$$e = e_0 := (0, 1, 0), \qquad m = m_0 := (0, 0, 1).$$

Similarly to [15], a covariant realisation is then obtained by a direct integral construction:

$$X := \int_G^{\oplus} \tilde{X}^{(\boldsymbol{a},R)}\, d\boldsymbol{a}dR, \tag{59}$$

where $X = \tau, \boldsymbol{x}, S, e_0\mathbb{I}, m_0\mathbb{I}$, and where of course $U(\boldsymbol{a}, R) = \mathbb{I} \otimes \lambda(\boldsymbol{a}, R)$ on $\int_G^{\oplus} L^2(\mathbb{R}^2)\, d\boldsymbol{a}dR \cong L^2(\mathbb{R}^2) \otimes L^2(G)$ and λ is the left regular representation of G. The construction from D of the corresponding domain $D^{(\boldsymbol{a},R)}$ such that the assumptions of Remark 4.1 are verified is now obvious.

Before concluding this work, we would like to point out that, following the lines of [31], it is possible to provide a completely different construction of the operators X above. In this case the main requirement is that $\mathrm{sp}(\tau)$ should coincide with the range of conformal time in the classical limit and then we build τ simply as the obvious multiplication operator on $L^2(\mathrm{sp}(\tau))$. Then, we set $\tilde{\beta e}\cdot x = [F(\tau), i\partial/\partial\tau]_+$. The problem here is that in general this operator is only maximally symmetric, a fact that forces us to use a certain "doubling construction" to obtain a selfadjoint operator (see [31]).

As a matter of fact, both the procedures outlined in this section lead to difficulties in the definition of a C^*-algebra corresponding to our non commutative Friedman spacetime. The way out will be presented in [26].

[1] Connes A, *Non commutative geometry*, A.P., 1994.
[2] Dixmier J, *Von Neumann algebras*, North-Holland, 1981.
[3] Hawking S, Ellis G *The large scale structure of space-time*, Cambridge University Press, 1975.
[4] Pedersen G K, *Analysis Now*, Springer 1989.
[5] Reed M, Simon B *Methods of modern mathematical physics*, Academic Press, 1980.
[6] Sakai S *C^*-algebras and W^*-algebras*, Springer, 1971.
[7] Schmüdgen K, *Unbounded operator algebras and representation theory*, Birkhäuser, 1990.
[8] Thiemann T, *Modern Canonical Quantum General Relativity*, Cambridge Univ. Press, 2007.
[9] Bahns D, Doplicher S, Fredenhagen K, Piacitelli G, Quantum Geometry on Quantum Spacetime: Distance, Area and Volume Operators, *Comm. Math. Phys.* **308** (2011), 567.
[10] Bahns D, Doplicher S, Fredenhagen, Piacitelli G, Ultraviolet finite quantum field theory on quantum spacetime, *Comm. Math. Phys.* **237** (2003), 221.
[11] Bahns D, Doplicher S, Fredenhagen K, Piacitelli G, On the unitarity problem in spacetime non commutative theories, *Phys.Lett. B* **533** (2002), 178.
[12] Bahns D, Doplicher S, Morsella G, Piacitelli G, Quantum Spacetime and Algebraic Quantum Field Theory, preprint, arXiv:1501.03298.
[13] Bizon P and Malec E, Trapped surfaces in spherical stars, *Phys. Rev. Lett.* (1988) **61**, 1147.
[14] Brauer U, Malec E, Murchadha N O, Trapped surfaces in spherical expanding open universes, *Phys. Rev. D*, (1994) **49**, 5601.
[15] Doplicher S, Fredenhagen K, Roberts J E, The quantum structure of spacetime at the Planck scale and quantum fields, *Commun. Math. Phys.* **172** (1995), 187.
[16] Doplicher S, Quantum Spacetime, *Ann. Inst. H. Poincaré Phys. Théor.* **64** (1996), 543.
[17] Doplicher S, Spacetime and Fields, a Quantum Texture, *Proceedings of the 37th Karpacz Winter School of Theoretical Physics*, (2001), 204.
[18] Doplicher S, Morsella G, Pinamonti N, On quantum spacetime and the horizon problem, *J. Geom. Phys.* **74** (2013), 196.
[19] Doplicher S, Piacitelli G, Tomassini L, Viaggiu S, Comment on *Can we measure structures to a precision better than the Planck length?*, by Sabine Hossenfelder, preprint, arXiv:1206.3067.
[20] Gracia-Bondía J M, Várilly J C, Algebras of distributions suitable for phase-space quantum mechanics. I, *J. Math. Phys.* **29** (1988), 869.
[21] Koc P, Malec E, Trapped Surfaces in Non Spherical Open Universes, *Acta Phys. Pol. B* **23** (1992), 123.
[22] Mars M, Present status of the Penrose inequality, *Class. Quantum Grav.* **19** (2009), 59.
[23] Martinetti P, Mercati F, and Tomassini L, Minimal length in quantum space and integrations of the line element in noncommutative geometry, *Rev. Math. Phys.* **24** (2012), 5.
[24] Martinetti P. and Tomassini L. , Length and distance on a quantum space, *Proc. of Sciences* **042** (2011).
[25] Mead C, Possible Connection between Gravitation and Fundamental Length, *Phys. Rev. B* **135** (1964), 849.
[26] Morsella G, Tomassini L, in preparation.
[27] Perini C, Tornetta G. N, A scale covariant quantum space-time, *Rev. Math. Phys.* **26** (2014), 112.
[28] Piacitelli G, Quantum Spacetime: a Disambiguation, *SIGMA* **6** (2010), 43.
[29] Piacitelli G, Twisted Covariance as a Non Invariant Restriction of the Fully Covariant DFR Model, arXiv:0902.0575 [hep-th].
[30] Tomassini L, Viaggiu S, Physically motivated uncertainty relations at the Planck length for an emergent non-commutative spacetime, *Class. Quantum Grav.* **28** (2011), 1.
[31] Tomassini L, Viaggiu S, Building non commutative spacetimes at the Planck length for Friedmann flat cosmologies, *Class. Quantum Grav.* **31** (2014), 1.

Phenomenology from the DSR-deformed relativistic symmetries of 3D quantum gravity via the relative-locality framework

Stefano Bianco

Max Planck Institute for Gravitational Physics (Albert Einstein Institute), Am Mühlenberg 1, D-14476 Potsdam, Germany

E-mail: stefanobianco3@gmail.com

Abstract. During the last decade, there have been relevant advances in the study of quantum gravity coupled to point particles in 2+1 dimensions. In the emerging picture the momentum space of the particles is curved, the spacetime coordinates are noncommuting and the symmetries are DSR-deformed relativistic symmetries. In this article we study some phenomenological consequences of these features via the so-called "relative-locality" framework. We find that a "dual-gravity" lensing effect appears as a consequence of the relativity of spacetime locality associated to the deformed symmetries of the theory.

1. Introduction

The quantum properties of gravity are still elusive because of theoretical and experimental challenges.

From the theoretical point of view, the quantum-field-theory techniques that are very successful for phenomena described by electro-weak and strong interactions lead to severe conceptual and technical difficulties when applied to gravity. Also the brand new approaches that have being developed have reached some partial results, but they still face key open problems.

From the experimental side, the energy scale at which quantum gravity is expected to become relevant is given by the Planck mass $M_p = 10^{19} Gev$, which is very distant from the energy scales that we can reach in current and near-future laboratory experiments. This means that in order to probe the quantum properties of gravity we need some amplification mechanisms that let the minute Planck-scale effects manifest.

The severe technical difficulties that one encounters when trying to formulate a full-comprehensive quantum theory of gravity have led to the extensive study of simpler systems that can provide theoretical and experimental insights into the quantum-gravity realm. In the last decade, there has been a significant progress in the quantum description of gravity coupled to point particles in 2+1 dimensions [1, 2, 3, 4, 5]. In particular, several approaches have led to the same class of results, which can now be considered as robust. Among these results there is the fact that in 2+1 dimensions and with vanishing cosmological constant the momentum space of a particle is described by the Lie group $SL(2,R)$, which has a curved (anti-deSitter) geometry. The departure from the flat (minkowskian) momentum space is controlled by the gravitational constant G, which provides an energy scale in 2+1 dimensions since its dimension is of the inverse of an energy. As a closely related feature one finds that the spacetime coordinates are characterized by a noncommutativity of the type $[x^\mu, x^\nu] = i\hbar\ell\varepsilon^{\mu\nu}{}_\rho x^\rho$, where by ℓ we denote the inverse of the energy scale given by G. These features result in a deformation in ℓ of the relativistic symmetries, which are no more described by the Poincaré group, but by the quantum group $\mathcal{D}SO(2,1)$.

It is interesting to note that the Planck-scale deformation of momentum space was suggested by an early argument by Born [6] and it is now supported by calculations in several quantum-gravity approaches in 3+1 dimensions, such as loop quantum gravity [7], group field theory [8], noncommutative geometry [9] and string theory [10].

It is important to remark that in the recent years there have been extensive studies on DSR-deformed special relativistic theories [11, 12, 13, 15, 16, 17]. These are theories in which the special relativistic symmetries are present but deformed by an energy scale which is expected to be the Planck scale, so that they are relativistic theories with two invariant constants: the speed of light c and the Planck scale. This class of theories have been shown to provide testable predictions for an important sort of quantum-gravity experiments, *i.e.* the observation of high-energy particles from astrophysical sources [18, 19, 20, 21] and, very recently, for cosmology [22, 23] (for more phenomenology coming from the Planck-scale deformation of the Poincaré symmetries we refer to [24]).

Recently a framework which enables to derive a DSR-deformed relativistic theory from a Planck-scale-curved momentum space has being developed and it goes under the name "relative-locality" framework [25, 26]. As we will see in more detail below, in the relative-locality framework momentum space is characterized by a metric and an affine connection which define respectively the (possibly-deformed) on-shell relation and the (possibly-deformed) composition law of momenta [25, 26]. So that thank to this framework one can study the kinematics of particles characterized by a curved momentum space.

This proceeding reports the results of Ref. [27], which considers the above mentioned results obtained in 3D quantum gravity with vanishing cosmological constant and extracts from them phenomenological predictions via the relative-locality framework. In particular we will make use of the classical limit of the structures appearing in 3D quantum gravity, as a first but very rich step of the analysis. In fact while in this limit we will not have spacetime noncommutativity, the curvature of momentum space and the deformation of the symmetries will be present, since, as we already noticed, in 3D the energy scale is given by the inverse of G. We will show that a "dual-gravity" lensing appears as a consequence of the relativity of spacetime locality, which, as we will explain in more detail below, is an important feature of DSR-theories.

We will adopt units such that $c = 1$ and, as already said, ℓ will denote the inverse of the Planck scale in 3D. The antisymmetric tensor $\varepsilon_{\mu\nu\rho}$ is such that $\varepsilon_{012} = -1$ and indices are raised and lowered with $\eta_{\mu\nu} = (-1,1,1)$.

2. Momentum space emerging from 3D quantum gravity

We here look in more detail at the momentum space and the spacetime noncommutativity that arise in 3D quantum gravity in the case of vanishing cosmological constant. As we have already said in the introduction the momentum space is described by the Lie group $SL(2,R)$. We can easily see that this momentum space is characterized by an anti-de-Sitter geometry. In fact we can write a generic element $\mathbf{p}$ of $SL(2,R)$ as a combination of the identity matrix and of the elements of a basis of $sl(2,R)$

$$\mathbf{p} = u\mathbb{I} - 2\xi_\mu X^\mu, \tag{1}$$

where $\mathbb{I}$ is the identity 2×2 matrix and the X^μ are

$$X^0 = \frac{1}{2}\begin{pmatrix} 0 & 1 \\ -1 & 0 \end{pmatrix}, X^1 = \frac{1}{2}\begin{pmatrix} 0 & -1 \\ -1 & 0 \end{pmatrix}, X^2 = \frac{1}{2}\begin{pmatrix} -1 & 0 \\ 0 & 1 \end{pmatrix}, \tag{2}$$

which is a basis of $sl(2,R)$. The requirement of having determinant equal to one ($\det\mathbf{p} = 1$) implies that the parameters (u, ξ_μ) satisfy

$$u^2 - \xi^\mu\xi_\mu = 1\,, \tag{3}$$

which is the definition of a 3 dimensional anti-de-Sitter hypersurface.

We notice that (2) implies that the X^μ satisfy (up to a dimensionful constant) the spacetime commutation relations that arise in 3D gravity

$$[X^\mu, X^\nu] = \varepsilon^{\mu\nu}{}_\rho X^\rho \ . \tag{4}$$

For our purposes we will adopt coordinates p_μ on our momentum space[1] such that

$$\mathbf{p} = \sqrt{1+\ell^2 p^\mu p_\mu}\mathbb{I} - 2\ell p_\mu X^\mu \ : \tag{5}$$

these p_μ will be the observable momenta. With this coordinatization of momentum space it has been shown [1, 2] that one finds a deformed onshell relation for the observbale momenta of the type

$$\ell^{-2}\left(\arcsin\left(\sqrt{-\ell^2 p^\mu p_\mu}\right)\right)^2 = m^2 \ , \tag{6}$$

where m is the mass of the particle. Moreover the group structure of momentum space implies a deformation of the composition law of momenta. In fact if we multiply two group elements $\mathbf{p}$ and $\mathbf{q}$

$$\begin{aligned} \mathbf{p} &= \sqrt{1+\ell^2 p^\mu p_\mu}\mathbb{I} - 2\ell p_\mu X^\mu \ , \\ \mathbf{q} &= \sqrt{1+\ell^2 q^\mu q_\mu}\mathbb{I} - 2\ell q_\mu X^\mu \ , \end{aligned} \tag{7}$$

we obtain a new element $\mathbf{pq}$

$$\mathbf{pq} = \left(\sqrt{1+\ell^2 p^\mu p_\mu}\sqrt{1+\ell^2 q^\nu q_\nu} + \ell^2 p^\mu q_\mu\right)\mathbb{I} - 2\ell\left(\sqrt{1+\ell^2 q^\nu q_\nu}p_\mu + \sqrt{1+\ell^2 p^\nu p_\nu}q_\mu - \ell\varepsilon_\mu{}^{\nu\rho}p_\nu q_\rho\right)X^\mu, \tag{8}$$

where we made use of the identity

$$X^\mu X^\nu = \frac{1}{4}\eta^{\mu\nu}\mathbb{I} + \frac{1}{2}\varepsilon^{\mu\nu}{}_\rho X^\rho \ . \tag{9}$$

From (8) we notice that the coordinates $(p\oplus q)_\mu$ of $\mathbf{pq}$ are related to the coordinates of p_μ, q_μ of $\mathbf{p}$,$\mathbf{q}$ in the following way

$$(p\oplus q)_\mu = \sqrt{1+\ell^2 q^\nu q_\nu}p_\mu + \sqrt{1+\ell^2 p^\nu p_\nu}q_\mu - \ell\varepsilon_\mu{}^{\nu\rho}p_\nu q_\rho \ : \tag{10}$$

this is the deformed composition law for the momenta p_μ, q_μ.

3. Deformed relativistic kinematics

We here describe the DSR-relativistic symmetries, that arise of 3D quantum gravity, in the classical limit, which is the regime we will focus on in this article. These will be compatible with the spacetime coordinates with Poisson brackets given by

$$\{x^\mu, x^\nu\} = \ell\varepsilon^{\mu\nu}{}_\rho x^\rho \ , \tag{11}$$

and by a momentum space with coordinates p_μ satisfying the onshell relation

$$\ell^{-2}\left(\arcsin\left(\sqrt{-\ell^2 p^\mu p_\mu}\right)\right)^2 = m^2 \ , \tag{12}$$

[1] An alternative coordinatization is reported in Ref. [1]). It consists in coordinates $P_\alpha, P_\beta, P_\gamma$ which are essentially Euler angles and are connected to the coordinates of our Eq. (5) by the relations $p_0 = \ell^{-1}\sin(P_\alpha\ell)\cosh(P_\beta\ell)$, $p_1 = \ell^{-1}\cos(P_\gamma\ell)\sinh(P_\beta\ell)$, $p_2 = \ell^{-1}\sin(P_\gamma\ell)\sinh(P_\beta\ell)$.

and obeying the composition law

$$(p\oplus q)_\mu = \sqrt{1+\ell^2 q^\nu q_\nu}\,p_\mu + \sqrt{1+\ell^2 p^\nu p_\nu}\,q_\mu - \ell\varepsilon_\mu{}^{\nu\rho} p_\nu q_\rho\,. \tag{13}$$

It is easy to see that the action of Lorentz-sector generators on momenta remains undeformed. In fact the mass shell (12) is invariant and the composition law (13) is covariant when adopting the standard

$$\{R,p_0\}=0\,, \qquad \{N_1,p_0\}=p_1\,, \qquad \{N_2,p_0\}=p_2\,, \tag{14}$$
$$\{R,p_1\}=-p_2\,, \qquad \{N_1,p_1\}=p_0\,, \qquad \{N_2,p_1\}=0\,, \tag{15}$$
$$\{R,p_2\}=p_1\,, \qquad \{N_1,p_2\}=0\,, \qquad \{N_2,p_2\}=p_0\,, \tag{16}$$

where R is the generator of rotations, while N_1 an N_2 are the generators of boosts.

Now we study the action of the symmetry generators on the spacetime coordinates. We will show that adopting undeformed Poisson brackets among translation generators

$$\{p_\mu,p_\nu\}=0 \tag{17}$$

and undeformed action of rotations and boosts on the spacetime coordinates

$$\{R,x^0\}=0\,, \qquad \{N_1,x^0\}=-x^1\,, \qquad \{N_2,x^0\}=-x^2\,, \tag{18}$$
$$\{R,x^1\}=-x^2\,, \qquad \{N_1,x^1\}=-x^0\,, \qquad \{N_2,x^1\}=0\,, \tag{19}$$
$$\{R,x^2\}=x^1\,, \qquad \{N_1,x^2\}=0\,, \qquad \{N_2,x^2\}=-x^0\,, \tag{20}$$

the deformation will concern only the action of the translations generators on the spacetime coordinates in order to have the Jacobi identity satisfied. From now on our analysis will be mainly concetrated on results at leading order in ℓ, and we will start showing that there is a unique Poisson bracket of $\{p_\mu,x^\nu\}$, given by

$$\{p_\mu,x^\nu\}\simeq -\delta_\mu^\nu + \frac{\ell}{2}\varepsilon_\mu{}^{\nu\rho}p_\rho\,, \tag{21}$$

which satisfies all the Jacobi identities, taking into account (15)-(21). To do this we start considering the following general form for $\{p_\mu,x^\nu\}$ at leading order which gives the standard result in the limit $\ell\to 0$ limit

$$\{p_\mu,x^\nu\} = -\delta_\mu^\nu + \ell f_\mu{}^{\nu\rho}p_\rho\,. \tag{22}$$

We begin determining the parameters $f_\mu{}^{\nu\rho}$ by considering the Jacobi identity

$$\{p_\mu,\{x^\nu,x^\rho\}\}+\{x^\rho,\{p_\mu,x^\nu\}\}+\{x^\nu,\{x^\rho,p_\mu\}\}=0\,, \tag{23}$$

which turns out to imply

$$f_\mu{}^{\nu\rho} - f_\mu{}^{\nu\rho} = \varepsilon_\mu{}^{\nu\rho}. \tag{24}$$

This fixes the antisymmetric part of $f_\mu{}^{\nu\rho}$ in the indices $\nu\rho$ so that we can now write

$$f_\mu{}^{\nu\rho} = t_\mu{}^{\nu\rho} + \frac{1}{2}\varepsilon_\mu{}^{\nu\rho}, \tag{25}$$

where $t_\mu{}^{\nu\rho}$ must be symmetric in the indices $\nu\rho$. Then we consider the Jacobi identities for R,p_μ,x^ν, for N_1,p_μ,x^ν, and for N_2,p_μ,x^ν. These give the following equations

$$f_\mu{}^{\nu\rho}\varepsilon_{\gamma\rho}{}^\lambda + f_\rho{}^{\nu\lambda}\varepsilon_{\mu\gamma}{}^\rho + f_\mu{}^{\rho\lambda}\varepsilon_{\gamma\rho}{}^\nu = 0. \tag{26}$$

We rewrite these equations using (25) and we find that ${t_\mu}^{\nu\rho}$ must satisfy

$$ {t_\mu}^{\nu\rho}{\varepsilon_{\gamma\rho}}^{\lambda} + {t_\rho}^{\nu\lambda}{\varepsilon_{\mu\gamma}}^{\rho} + {t_\mu}^{\rho\lambda}{\varepsilon_{\gamma\rho}}^{\nu} = 0 \, . \tag{27} $$

Now using that ${t_\mu}^{\nu\rho}$ is symmetric for the exchange of indices $\nu\rho$, one finds

$$ {t_\mu}^{\nu\rho} = 0 \, . \tag{28} $$

So looking back at (25) we realize that the choice ${f_\mu}^{\nu\rho} = \frac{1}{2}{\varepsilon_\mu}^{\nu\rho}$ is unique. As an aside we note that the following all-orders-in-ℓ formula

$$ \{p_\mu, x^\nu\} = -\delta^\nu_\mu \sqrt{1 + \frac{\ell^2}{4} p^\rho p_\rho} + \frac{\ell}{2}{\varepsilon_\mu}^{\nu\rho} p_\rho \tag{29} $$

satisfies all the Jacobi identities exactly (no leading-order truncation), taking into account $\{x^\mu, x^\nu\} = \ell {\varepsilon^{\mu\nu}}_\rho x^\rho$ and (15)-(21).

We note that there is another source of deformation of the translational symmetry, which appears in multiparticle systems. This deformation is due to the fact that the momentum charges are composed following the deformed addition law (13). In fact if we consider for example a system of two particles, with phase-space coordinates p_μ, x^ν and q_μ, y^ν, a translation with parameter b^ρ is generated by the total-momentum charge $(p \oplus q)_\rho$. For the particle with phase-space coordinates p_μ, x^ν this translation transformation results in

$$ b^\rho \{(p \oplus q)_\rho, x^\nu\} \simeq b^\rho \{p_\rho, x^\nu\} - \ell b^\rho {\varepsilon_\rho}^{\sigma\gamma} q_\gamma \{p_\sigma, x^\nu\} \, , \tag{30} $$

which is a deformation of the usual translation due to the nonlinearity of the composition law and it can be described by saying that the symmetries have the structure of a quantum group with nontrivial coproduct for the translation generators.

We have thus characterized the DSR-symmetries that arise in 3D quantum gravity in the classical limit and we have seen that they are such that the action of Lorentz-sector generators is undeformed while the action of translation generators is deformed.

4. Kinenamics within the relative-locality framework

In this section we will first show how the onshell relation and the composition law of momenta can be interpreted with the relative-locality framework in terms of the geometric properties of the momentum space. Then we will see how using this framework one is able to describe the classical relativistic kinematics of a system of particles characterized by the momentum space $SL(2,R)$ which emerges from 3D gravity.

4.1. On-shell relation from the relative-locality framework

In the relative-locality framework [25, 26], the onshell relation is encoded in the metric on momentum space. In particular it is defined by the geodesic distance $D(0, p_\mu)$ of the momentum p_μ from the origin:

$$ D^2(0, p_\mu) = m^2 \, , \tag{31} $$

where $D(0, p_\mu) = \int_0^1 g^{\mu\nu} \dot{k}_\mu(s) \dot{k}_\mu(s) ds$, $k_\mu(0) = 0$, $k_\mu(1) = p_\mu$ and the geodesic is defined by the Levi-Civita connection associated to the momentum space metric $g^{\mu\nu}$.

We now find a metric on our momentum space $SL(2,R)$ that encodes the onshell relation (42). To do this we embed $SL(2,R)$ in $\mathbb{R}^{2,2}$, which is characterized by the metric

$$ ds^2 = -du^2 - (d\xi_0)^2 + (d\xi_1)^2 + (d\xi_2)^2 \, . \tag{32} $$

Journal of Physics: Conference Series **634** (2015) 012003 doi:10.1088/1742-6596/634/1/012003

So we can describe the metric on $SL(2,R)$ as the metric induced by the $\mathbb{R}^{2,2}$ metric. The embedding coordinates are $Y_I = (\sqrt{1+\ell^2 p^\mu p_\mu}, \ell p_\mu)$ so that the pull-back of the metric (32) to our $SL(2,R)$ gives us[2]:

$$ds^2 = -(dp_0)^2 + (dp_1)^2 + (dp_2)^2 - \frac{\ell^2 p^\mu p^\nu dp_\mu dp_\nu}{1+\ell^2 p^\mu p_\mu}\,, \tag{33}$$

which will be the metric we adopt on $SL(2,R)$.

Our next step is to compute $D(0,p_\mu)$ using the metric (33) and to show that we are able to reproduce the onshell relation (42). For our purpose it is convenient to describe the geodesics in the embedding space $\mathbb{R}^{2,2}$. And we start noticing that on the anti-deSitter hypersurface, which is the image of our embedding, a geodesic defined by the Levi-Civita connection associated to the metric (33) can be described by the Lagrangian

$$L = \dot{Y}^I \dot{Y}_I + \lambda(Y^I Y_I + 1)\,, \tag{34}$$

where λ is a Lagrange multiplier, imposing that the motion should be on the anti-deSitter hypersurface. We note that the equations of motion coming from the Lagrangian (34) are very simple:

$$\begin{aligned} \ddot{Y}_I &= \lambda Y_I \\ Y^I Y_I + 1 &= 0. \end{aligned} \tag{35}$$

It is easy to see that when computing the geodesic distance we can distinguish three cases. In fact for the geodesic going out from the origin and arriving at a point $Y_I = (\sqrt{1+\ell^2 p^\mu p_\mu}, \ell p_\mu)$ we have

$$\ell^2 p^\mu p_\mu = \sinh^2(D(0,Y_I))\,, \tag{36}$$

if $p^\mu p_\mu > 0$. If $p^\mu p_\mu = 0$ one gets $D(0,Y_I) = 0$ and finally, if $p^\mu p_\mu < 0$ one has

$$\ell^2 p^\mu p_\mu = -\sin^2(D(0,Y_I)). \tag{37}$$

Using that $D(0,Y_I) = \ell D(0,p_\mu) = \ell m$ we rewrite the previous equations as

$$\ell^2 p^\mu p_\mu = \sinh^2(\ell m) \qquad p^\mu p_\mu > 0\,, \tag{38}$$

$$p^\mu p_\mu = 0 \qquad p^\mu p_\mu = 0\,, \tag{39}$$

$$\ell^2 p^\mu p_\mu = -\sin^2(\ell m) \qquad p^\mu p_\mu < 0\,. \tag{40}$$

Since our mass-shell condition should be a perturbation of the special relativistic one, the relevant cases are the last two, that can be written together as

$$\ell^2 p^\mu p_\mu = -\sin^2(\ell m)\,. \tag{41}$$

Rewriting (41) in the spirit of (31), we finally have

$$\ell^{-2}\left(\arcsin\left(\sqrt{-\ell^2 p^\mu p_\mu}\right)\right)^2 = m^2\,, \tag{42}$$

which reproduces the prediction (42) based on 3D-gravity results.

We note that the physical momentum space is defined by the condition

$$-\ell^{-2} \le p^\mu p_\mu \le 0, \tag{43}$$

where the first inequality comes from the anti-deSitter nature of our momentum space and the second one comes from the requirement for the mass-shell condition to have the right special relativistic limit.

[2] The metric (33) on $SL(2,R)$ can also be obtained by exploiting the fact that our momentum space is a Lie group and then defining the metric over this space using the Killing form of its Lie algebra.

4.2. Composition law of momenta from the relative-locality framework

In the relative-locality framework [25, 26], the law of composition of momenta is encoded in the affine connection on momentum space, which is in principle not related to the metric which gives the onshell relation. In the leading-order analysis we are interested in here, the momentum-space connection coefficients $\Gamma_\mu^{\nu\rho}(0)$ are the coefficients of the leading-order expansion of the composition law [25, 26]:

$$(p\oplus q)_\mu \simeq p_\mu + q_\mu - \ell\Gamma_\mu{}^{\nu\rho}(0)p_\nu q_\rho + \dots \tag{44}$$

Recalling that in our case we have

$$(p\oplus q)_\mu \simeq p_\mu + q_\mu - \ell\varepsilon_\mu{}^{\nu\rho}p_\nu q_\rho \;. \tag{45}$$

then the connection coefficients which encode the composition law in our case are

$$\Gamma_\mu^{\nu\rho}(0) = \varepsilon_\mu{}^{\nu\rho} \;. \tag{46}$$

Adopting this geometric description, the algebraic properties of the composition law become geometric properties, which can be analyzed through the computation of the torsion, curvature and nonmetricity tensors. Following the definitions given in Refs. [25, 26], the torsion describes the noncommutativity of the composition law and we find

$$T_\mu^{\nu\rho}(0) = -\frac{\partial}{\partial p_\nu}\frac{\partial}{\partial q_\beta}((p\oplus q)_\mu - (q\oplus p)_\mu)_{p=q=0} = 2\ell\Gamma_\mu^{[\nu\rho]}(0) = \ell\Gamma_\mu^{\nu\rho}(0) - \Gamma_\mu^{\rho\nu}(0) = 2\ell\varepsilon_\mu{}^{\nu\rho} \;. \tag{47}$$

The curvature of the connection (evaluated in the origin) measures the nonassociativity of the composition law and in our case it vanishes

$$R_\mu^{\nu\rho\sigma}(0) = 2\frac{\partial}{\partial p_{[\nu}}\frac{\partial}{\partial q_{\rho]}}\frac{\partial}{\partial k_\sigma}((p\oplus q)\oplus k - p\oplus(q\oplus k))_\mu\,|_{p=q=k=0} = 0 \;, \tag{48}$$

meaning that the composition law is associative. We finally compute the value in the origin of the nonmetricity tensor, which turns out to be 0:

$$N^{\rho\mu\nu}(0) = \nabla^\rho g^{\mu\nu}(0) = g^{\mu\nu,\rho}(0) + \ell\Gamma_\sigma^{\mu\rho}(0)g^{\sigma\nu}(0) + \ell\Gamma_\sigma^{\nu\rho}(0)g^{\mu\sigma}(0) = 0.$$

4.3. Description of particles within the relative-locality framework

We here show with a simple example how one describes the kinematics of a system of interacting particles characterized by a curved momentum space in the relative-locality framework. As we will see, the key ingredients Refs. [25, 26] are the deformed onshell relation and boundary terms, enforcing momentum conservation, at endpoints of wordlines, when an interaction occurs. For our analysis we will adopt the $SL(2,R)$ momentum space and the associated deformed relativistic kinematics described previously, and we will be satisfied by considering formulas at leading order in ℓ. In particular our example is a single two-body-particle-decay process (see Fig. 1), and at leading order in ℓ with the prescription given above it is described by the action [25, 26]:

$$\begin{aligned} \mathcal{S} \;=\; & \int_{-\infty}^{s_0}((\delta^\mu_\nu - \frac{\ell}{2}\varepsilon^{\mu\sigma}{}_\nu k_\sigma)z^\nu\dot{k}_\mu + \mathcal{N}_k(k^\mu k_\mu - m^2))ds + \int_{s_0}^{\infty}((\delta^\mu_\nu - \frac{\ell}{2}\varepsilon^{\mu\sigma}{}_\nu p_\sigma)x^\nu\dot{p}_\mu + \mathcal{N}_p(p^\mu p_\mu - m'^2))ds \\ & + \int_{s_0}^{\infty}((\delta^\mu_\nu - \frac{\ell}{2}\varepsilon^{\mu\sigma}{}_\nu q_\sigma)y^\nu\dot{q}_\mu + \mathcal{N}_q(q^\mu q_\mu - m'^2))ds - \xi^\mu_{[0]}\mathcal{K}^{[0]}_\mu(s_0)\;. \end{aligned} \tag{49}$$

Here the Lagrange multipliers $\mathcal{N}_k, \mathcal{N}_p, \mathcal{N}_q$ enforce the on-shell relation of particles, m represents the mass of the incoming particle and m' and m'' the masses of the outgoing particles. We note that the symplectic form is such that $\{x^\mu, x^\nu\} = \ell\varepsilon^{\mu\nu}{}_\rho x^\rho$, so that (see (29)) one has

$$\{p_\mu, x^\nu\} = -\delta^\nu_\mu + \frac{\ell}{2}\varepsilon_\mu{}^{\nu\sigma} p_\sigma \,. \tag{50}$$

$\xi^\mu_{[0]}$ is a Lagrange multiplier enforcing the conservation law of momenta, and, on the basis of what we have established in the previous sections, we have

$$\mathcal{K}^{[0]}_\mu(s_0) = (k)_\mu - (p \oplus q)_\mu = k_\mu - p_\mu - q_\mu + \ell\varepsilon_\mu{}^{\nu\rho} p_\nu q_\rho \,. \tag{51}$$

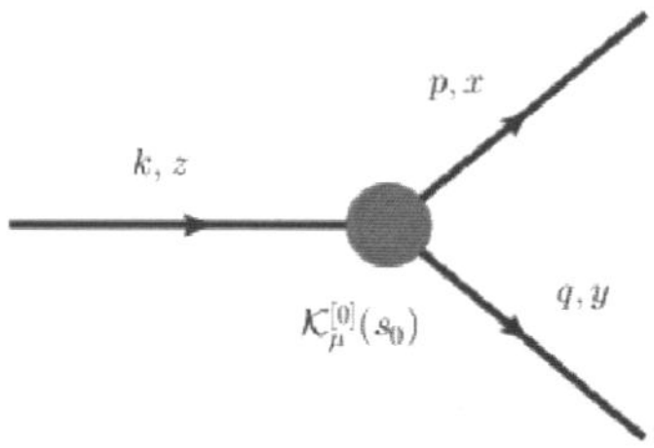

Figure 1. We here show the decay process described in Eq. (49), with one incoming and two outgoing particles.

It has been noticed [25, 26, 28] that the deformation of the relativistic symmetries produces relativity of spacetime locality[3]. This amounts to the fact that an interaction event which is local in the description of nearby observers, is not local in the description of distant observers. And this feature leads to the conclusion that the description of an event made by distant observers is misleading and that one should use the description of an event made by an observer which is near to it. We will see in the next section that the relativity of spacetime locality will require to use both equations of motion and symmetry transformations when deriving physical predictions from the theory.

5. Dual-gravity lensing

We here analyze a phenomenological consequence deriving from adopting the $SL(2,R)$ momentum space and associated deformed relativistic kinematics described in the previous sections. To do this we will consider the process depicted in Fig. 2, which is described within the relative-locality framework by the following action

$$\begin{aligned} \mathcal{S} = & \int_{-\infty}^{s_0} ((\delta^\mu_\nu - \frac{\ell}{2}\varepsilon^{\mu\sigma}{}_\nu p_\sigma) x^\nu \dot{p}_\mu + \mathcal{N}_p(p^\mu p_\mu - m^2)) ds + \int_{s_0}^{+\infty} ((\delta^\mu_\nu - \frac{\ell}{2}\varepsilon^{\mu\sigma}{}_\nu k_\sigma) z^\nu \dot{k}_\mu + \mathcal{N}_k(k^\mu k_\mu \\ & - m'^2)) ds + \int_{s_0}^{s_1} ((\delta^\mu_\nu - \frac{\ell}{2}\varepsilon^{\mu\sigma}{}_\nu p'_\sigma) x'^\nu \dot{p}'_\mu + \mathcal{N}_{p'}(p'^\mu p'_\mu - m''^2)) ds + \int_{-\infty}^{s_1} ((\delta^\mu_\nu - \frac{\ell}{2}\varepsilon^{\mu\sigma}{}_\nu q_\sigma) y^\nu \dot{q}_\mu \\ & + \mathcal{N}_q(q^\mu q_\mu - \mu^2)) ds + \int_{s_1}^{+\infty} ((\delta^\mu_\nu - \frac{\ell}{2}\varepsilon^{\mu\sigma}{}_\nu p''_\sigma) x''^\nu \dot{p}''_\mu + \mathcal{N}_{p''}(p''^\mu p''_\mu) - \mu'^2)) ds + \int_{s_1}^{+\infty} ((\delta^\mu_\nu \\ & - \frac{\ell}{2}\varepsilon^{\mu\sigma}{}_\nu q'_\sigma) y'^\nu \dot{q}'_\mu + \mathcal{N}_{q'}(q'^\mu q'_\mu - \mu''^2)) ds - \xi^\mu_{(0)} \mathcal{K}^{[0]}_\mu(s_0) - \xi^\mu_{(1)} \mathcal{K}^{[1]}_\mu(s_1) \,, \end{aligned} \tag{52}$$

[3] The relativity of spacetime locality can also be seen as a consequence of the addtional invariant (energy) scale that is present in DSR theories. In fact, in a similar way, in the passage from galilean to special relativity, the introduction of the invariant constant c, renders simultaneity an observer dependent feature.

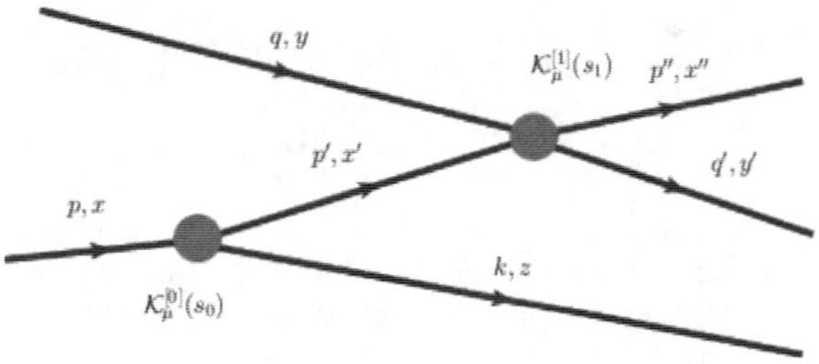

Figure 2. We here show the process described in Eq. (52), involving two causally-connected interactions.

where $(m,m',m'',\mu,\mu',\mu'')$ are the masses of the particles. The conservation laws are encoded in $\mathcal{K}_\mu^{[0]}(s_0)$ and $\mathcal{K}_\mu^{[1]}(s_1)$, for which we follow the prescription given in Ref. [26], for having causally-connected interactions preserving translational invariance, so that we have

$$\mathcal{K}_\mu^{[0]}(s_0) = (q\oplus p)_\mu - (q\oplus p'\oplus k)_\mu = p_\mu - p'_\mu - k_\mu - \ell\varepsilon_\mu{}^{\alpha\beta}(q_\alpha p_\beta - q_\alpha p'_\beta - q_\alpha k_\beta - p'_\alpha k_\beta) \tag{53}$$

and

$$\begin{aligned}\mathcal{K}_\mu^{[1]}(s_1) = (q\oplus p'\oplus k)_\mu - (p''\oplus q'\oplus k)_\mu \;&=\; q_\mu + p'_\mu - p''_\mu - q'_\mu - \ell\varepsilon_\mu{}^{\alpha\beta}(q_\alpha p'_\beta + q_\alpha k_\beta + p'_\alpha k_\beta - p''_\alpha q'_\beta \\ &\quad - p''_\alpha k_\beta - q'_\alpha k_\beta)\,.\end{aligned} \tag{54}$$

The equations of motion and the boundary conditions at endpoints of worldlines (which come by the presence of the boundary terms devoted to enforce the conservation laws for an interaction event) are derived by varying (52) keeping momenta fixed [25, 26] at $\pm\infty$. Then for the equations of motion one finds

$$\dot{p}_\mu = 0\,,\;\; \dot{q}_\mu = 0\,,\;\; \dot{q}'_\mu = 0\,,\;\; \dot{k}_\mu = 0\,,\;\; \dot{p}'_\mu = 0\,,\;\; \dot{p}''_\mu = 0\,, \tag{55}$$

$$\mathcal{C}_p = 0\,,\;\; \mathcal{C}_q = 0\,,\;\; \mathcal{C}_{q'} = 0\,,\;\; \mathcal{C}_k = 0\,,\;\; \mathcal{C}_{p'} = 0\,,\;\; \mathcal{C}_{p''} = 0\,, \tag{56}$$

$$\dot{x}^\mu = 2\mathcal{N}_p p^\mu\,,\;\; \dot{y}^\mu = 2\mathcal{N}_q q^\mu\,,\;\; \dot{y}'^\mu = 2\mathcal{N}_{q'} q'^\mu\,, \tag{57}$$

$$\dot{z}^\mu = 2\mathcal{N}_k k^\mu\,,\;\; \dot{x}'^\mu = 2\mathcal{N}_{p'} p'^\mu\,,\;\; \dot{x}''^\mu = 2\mathcal{N}_{p''} p''^\mu\,, \tag{58}$$

and the boundary conditions are

$$\begin{aligned}
z^\mu(s_0) &= -\xi^\nu_{[0]}\frac{\delta\mathcal{K}_\nu^{[0]}}{\delta k_\sigma}(\delta^\mu_\sigma + \frac{\ell}{2}\varepsilon^\mu{}_\sigma{}^\rho k_\rho) = \xi^\mu_{[0]} - \frac{\ell}{2}\varepsilon_\nu{}^{\mu\alpha}(k_\alpha - 2q_\alpha - 2p'_\alpha)\xi^\nu_{[0]}\,,\\
x^\mu(s_0) &= \xi^\nu_{(0)}\frac{\delta\mathcal{K}_\nu^{(0)}}{\delta p_\sigma}(\delta^\mu_\sigma + \frac{\ell}{2}\varepsilon^\mu{}_\sigma{}^\rho p_\rho) = \xi^\mu_{[0]} - \frac{\ell}{2}\varepsilon_\nu{}^{\mu\alpha}(p_\alpha - 2q_\alpha)\xi^\nu_{[0]}\,,\\
x'^\mu(s_0) &= -\xi^\nu_{[0]}\frac{\delta\mathcal{K}_\nu^{[0]}}{\delta p'_\sigma}(\delta^\mu_\sigma + \frac{\ell}{2}\varepsilon^\mu{}_\sigma{}^\rho p'_\rho) = \xi^\mu_{[0]} - \frac{\ell}{2}\varepsilon_\nu{}^{\mu\alpha}(p'_\alpha - 2k_\alpha + 2q_\alpha)\xi^\nu_{[0]}\,,\\
x'^\mu(s_1) &= \xi^\nu_{[1]}\frac{\delta\mathcal{K}_\nu^{[1]}}{\delta p'_\sigma}(\delta^\mu_\sigma + \frac{\ell}{2}\varepsilon^\mu{}_\sigma{}^\rho p'_\rho) = \xi^\mu_{[1]} - \frac{\ell}{2}\varepsilon_\nu{}^{\mu\alpha}(p'_\alpha - 2k_\alpha + 2q_\alpha)\xi^\nu_{[1]}\,,\\
x''^\mu(s_1) &= -\xi^\nu_{[1]}\frac{\delta\mathcal{K}_\nu^{[1]}}{\delta p''_\sigma}(\delta^\mu_\sigma + \frac{\ell}{2}\varepsilon^\mu{}_\sigma{}^\rho p''_\rho) = \xi^\mu_{[1]} - \frac{\ell}{2}\varepsilon_\nu{}^{\mu\alpha}(p''_\alpha - 2q'_\alpha - 2k_\alpha)\xi^\nu_{[1]}\,,\\
y^\mu(s_1) &= \xi^\nu_{[1]}\frac{\delta\mathcal{K}_\nu^{[1]}}{\delta q_\mu}(\delta^\mu_\sigma + \frac{\ell}{2}\varepsilon^\mu{}_\sigma{}^\rho q_\rho) = \xi^\mu_{[1]} - \frac{\ell}{2}\varepsilon_\nu{}^{\mu\alpha}(q_\alpha - 2p'_\alpha - 2k_\alpha)\xi^\nu_{[1]}\,,
\end{aligned} \tag{59}$$

$$
\begin{aligned}
y'^{\mu}(s_1) &= -\xi^{\nu}_{[1]}\frac{\delta\mathcal{K}^{[1]}_{\nu}}{\delta q'_{\sigma}}(\delta^{\mu}_{\sigma}+\frac{\ell}{2}\varepsilon^{\mu}{}_{\sigma}{}^{\rho}q'_{\rho}) = \xi^{\mu}_{[1]} - \frac{\ell}{2}\varepsilon_{\nu}{}^{\mu\alpha}(q'_{\alpha}-2k_{\alpha}+2p''_{\alpha})\xi^{\nu}_{[0]}\,, \\
\mathcal{K}^{[0]}_{\mu}(s_0) &= 0\,,\ \ \mathcal{K}^{[1]}_{\mu}(s_1) = 0\,, \frac{\delta\mathcal{K}^{[0]}_{\mu}}{\delta q_{\nu}} = 0\,,\ \ \frac{\delta\mathcal{K}^{[1]}_{\mu}}{\delta k_{\nu}} = 0\,.
\end{aligned}
\tag{60}
$$

We note that when all the particles are "soft", *i.e.* they have energies small enough that the ℓ-deformation can be ignored, one finds the standard special-relativistic situation.

Here we will look in particular at what the theory predicts for the particle exchanged between the two interactions, assuming it is a massless particle. To do this we will consider an observer Alice who is placed where the interaction characterized by the conservation law $\mathcal{K}^{[0]}_{\mu} = 0$ occurs, and an observer Bob who is placed where the interaction with conservation law $\mathcal{K}^{[1]}_{\mu} = 0$ occurs. And for our purposes we will assume that Bob is at rest with respect to Alice and that Bob's and Alice's coordinatizations are related by a translation along the x^1 axis. We start analyzing what happens in the case in which all particles are soft, which gives the special-relativistic situation. In this case Alice's description of the exchanged massless particle is

$$
x'^{1}_{A(s)} = x'^{0}_{A(s)}\,,\quad x'^{2}_{A(s)} = 0\,,
\tag{61}
$$

where we assumed that the first interaction happens in Alice's origin, and that the massless particle exchanged between the two interactions propagates along the x^1 axis in Alice's coordinatization. The index (s) introduced in (61) will be used to identify quantities related to a soft particle. Now we require that the massless particle exchanged between the two interactions is detected, through the second interaction, in Bob's origin. This implies that Bob is related to Alice by a translation of parameters $b^{\mu} = (b^0, b^1, 0)$ with $b^0 = b^1$, and it is easy to see that the worldline in Bob's description is

$$
x'^{1}_{B(s)} = x'^{0}_{B(s)} + b^0 - b^1 = x'^{0}_{B(s)}\,,\quad x'^{2}_{B(s)} = 0\,.
\tag{62}
$$

which has the same form of the worldline in Alice's description.

Now we analyze what happens if the interactions are not soft and then the ℓ-corrections become important. For our analysis we choose the momentum of the exchanged particle to be (in Alice's and Bob's frames):

$$
\begin{aligned}
p'^{1} &= p'^{0}\cos\theta\,, \\
p'^{2} &= -p'^{0}\sin\theta\,,
\end{aligned}
\tag{63}
$$

where θ is a free parameter. Requiring that the particle is emitted from Alice's origin, one finds that the worldline in Alice's description is

$$
\begin{aligned}
x'^{1}_{A} &= x'^{0}_{A}\cos\theta\,, \\
x'^{2}_{A} &= -x'^{0}_{A}\sin\theta\,.
\end{aligned}
\tag{64}
$$

Now we want to check for which values of θ the exchanged particle crosses Bob's origin. For this, due to the relativity of spacetime locality mentioned in the previous paragraph, we need to determine Bob's description of the worldline of the exchanged particle. We start noting that the translation transformations which are symmetry of the action (52) are [25, 26]:

$$
\begin{aligned}
x'^{\mu}_{B} &= x'^{\mu}_{A} + b^{\nu}\{(q\oplus p'\oplus k)_{\nu}, x'^{\mu}\} = x'^{\mu}_{A} - b^{\mu} + b^{\nu}\frac{\ell}{2}\varepsilon_{\nu}{}^{\mu\rho}(p'_{\rho}+2k_{\rho}-2q_{\rho})\,, \\
x^{\mu}_{B} &= x^{\mu}_{A} + b^{\nu}\{(q\oplus p)_{\nu}, x^{\mu}_{A}\} = x^{\mu}_{A} - b^{\mu} + \frac{\ell}{2}\varepsilon_{\nu}{}^{\mu\rho}(p_{\rho}-2q_{\rho})b^{\nu}\,, \\
y^{\mu}_{B} &= y^{\mu}_{A} + b^{\nu}\{(q\oplus p'\oplus k)_{\nu}, y^{\mu}_{A}\} = y^{\mu}_{A} - b^{\mu} + \frac{\ell}{2}\varepsilon_{\nu}{}^{\mu\rho}(q_{\rho}+2p'_{\rho}+2k_{\rho})b^{\nu}\,, \\
z^{\mu}_{B} &= z^{\mu}_{A} + b^{\nu}\{(q\oplus p'\oplus k)_{\nu}, z^{\mu}_{A}\} = z^{\mu}_{A} - b^{\mu} + \frac{\ell}{2}\varepsilon_{\nu}{}^{\mu\rho}(k_{\rho}-2q_{\rho}-2p'_{\rho})b^{\nu}\,, \\
x''^{\mu}_{B} &= x''^{\mu}_{A} + b^{\nu}\{(p''\oplus q'\oplus k)_{\nu}, x''^{\mu}_{A}\} = x''^{\mu}_{A} - b^{\mu} + \frac{\ell}{2}\varepsilon_{\nu}{}^{\mu\rho}(p''_{\rho}+2q'_{\rho}+2k_{\rho})b^{\nu}\,, \\
y'^{\mu}_{B} &= y'^{\mu}_{A} + b^{\nu}\{(p''\oplus q'\oplus k)_{\nu}, y'^{\mu}_{A}\} = y'^{\mu}_{A} - b^{\mu} + \frac{\ell}{2}\varepsilon_{\nu}{}^{\mu\rho}(q'_{\rho}+2k_{\rho}-2p''_{\rho})b^{\nu}\,.
\end{aligned}
\tag{65}
$$

Then using the translation transformation for the coordinate of the exchanged particle, specializing it to the case of a translation from Alice to Bob for which the translation parameter is $b^\mu = (b, b, 0)$ and combining it with (64) , one finds that Bob's description of the worldline is

$$x_B'^1 = \cos\theta\, x_B'^0 + \cos\theta\Delta^0 - \Delta^1 \,, \tag{66}$$
$$x_B'^2 = -\sin\theta x_B'^0 - \sin\theta\Delta^0 - \Delta^2 \,, \tag{67}$$

where $\Delta^\mu \equiv b^\mu - \frac{\ell}{2}\varepsilon_\nu{}^{\mu\rho}(p'_\rho + 2k_\rho - 2q_\rho)b^\nu$. Now we observe that the equation of motion can be easily rearranged as follows

$$\begin{aligned} x_B'^2 &= -\tan\theta x_B'^1 - \tan\theta\Delta^1 - \Delta^2 \,, \\ x_B'^0 &= \frac{x_B'^1 - \cos\theta\Delta^0 + \Delta^1}{\cos\theta} \,. \end{aligned} \tag{68}$$

And finally enforcing that the particle goes through the space origin of Bob $x_B'^1 = x_B'^2 = 0$ we find that

$$\begin{aligned} \tan\theta &= -\frac{\Delta^2}{\Delta^1} \,, \\ x_B'^0 &= \frac{\Delta^1}{\cos\theta} - \Delta^0 \,. \end{aligned} \tag{69}$$

From the first equation we have

$$\theta \simeq \tan\theta = -\frac{\Delta^2}{\Delta^1} = -\frac{b\ell(k_1 - q_1 + k_0 - q_0) + b\frac{\ell}{2}(p'_0 + p'_1)}{-b - b\ell(k_2 - q_2) - b\frac{\ell}{2}p'_2} \approx \frac{b\ell(k_1 - q_1 + k_0 - q_0)}{b + b\ell(k_2 - q_2)} \simeq \ell(k_1 - q_1 + k_0 - q_0) \,, \tag{70}$$

which implies that the worldline of hard particle that reaches Bob from Alice is not parallel to the wordline of a soft particle that reaches Bob from Alice. This is the feature known as "dual-gravity lensing" in the relative-locality literature [29, 30]. Before commenting more on this effect, we now compute the time at which the hard particle crosses Bob's spatial origin. This can be done by using our result for θ in the second of Eqs. (69):

$$x_B'^0 = \frac{\Delta^1}{\cos\theta} - \Delta^0 \approx \Delta^1 - \Delta^0 \approx b + b(k_2 - q_2) - b - b(k_2 - q_2) = 0 \,. \tag{71}$$

So we have that the hard particle crosses the spacetime origin of Bob.

We make two further comments on the dual-gravity lensing we have found here and that we represent pictorially in Fig. 3. First we notice that, as in the previous studies which have considered this effect, relative locality plays a key role [29, 30]. In fact if we look at the event where the soft (red) worldline crosses Bob and the event where the hard (blue) worldline crosses Bob, we notice that these two events are coincident in the coordinatization of the nearby observer Bob, while in the description of the distant observer Alice they are not coincident. So in order to characterize the dual-gravity lensing it has been crucial not to rely only on the description of one observer, but to consider the description of the observer local to the event under consideration. Finally we make a remark on the energy dependence of the dual-gravity lensing we have found here. We notice that the angle θ goes like $\ell E_\star$, where $E_\star$ is the energy scale of the particles involved. Then considering a case with some $E_\star$ and a case with some $E'_\star$ bigger than $E_\star$ we have that the difference between the angles of the two cases is $\theta' - \theta \approx \ell E'_\star - \ell E_\star$. This contributes to the investigation of the energy dependence of the dual-gravity lensing. In fact, before our investigation, there have been the studies of [30] which predicts, as we do, $\theta' - \theta$ to be proportional to the difference of the energy scales involved, $E'_\star - E_\star$, while the study [29] predicts $\theta' - \theta$, to be proportional to the sum $E'_\star + E_\star$.

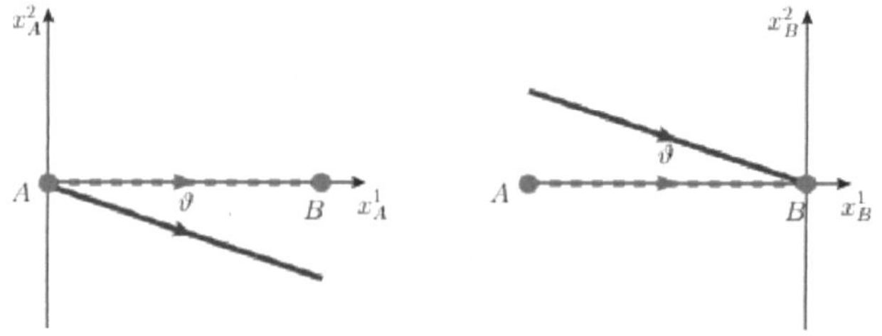

Figure 3. The worldline of a hard (high-energy) particle emitted at Alice and reaching Bob forms a non-zero angle with the wordline of a soft (low-energy) particle also emitted at Alice and reaching Bob. We note that the angle θ (whose value is determined by our Eq. (70)) is very small (for example if the energy of the particles are of the order of $1TeV$, then the angle θ is of order 10^{-16}.).

6. Outlook

We have seen how DSR-studies enhances the capacity of results from 3D quantum gravity, such as a curvature of momentum space, spacetime noncommutativity and deformed relativistic symmetries, to suggest interesting phenomenology. On the other hand 3D quantum gravity is helpful for DSR-research since DSR-models derived by 3D quantum gravity can be a guidance in the investigation of DSR-theories.

An interesting development of this study would be to investigate the phenomenology deriving by the quantum effects related to the deformation of the symmetries in 3D gravity. Similar studies have been performed only very recently in the context of κ-Minkowski noncommutativity [32, 33], and we feel that a good starting point to do this in the context of 3D gravity is constituted by the results in Ref. [31].

Acknowledgements

This study was done in collaboration with Giovanni Amelino-Camelia, Michele Arzano and Riccardo J. Buonocore. The writing of this article is supported by the Foundation Blanceflor Boncompagni-Ludovisi, neé Bildt.

References

[1] Matschull H J and Welling M 1998 Quantum mechanics of a point particle in (2+1)-dimensional gravity *Class. Quant. Grav.* **15** 2981–30 (*Preprint* gr-qc/9708054)
[2] Bais F A, Muller N M and Schroers B J 2002 Quantum group symmetry and particle scattering in (2+1)-dimensional quantum gravity *Nucl. Phys.* B**640** 3–45 (*Preprint* hep-th/0205021)
[3] Schroers B J 2011 Quantum gravity and non-commutative spacetimes in three dimensions: a unified approach *Acta Phys. Polon. Supp.* **4** 379 (*Preprint* 1105.3945 [gr-qc])
[4] Freidel L and Livine E R 2006 3d quantum gravity and effective non-commutative quantum field theory *Phys. Rev. Lett.* **96** 221301 (*Preprint* hep-th/0512113)
[5] Oriti D and Ryan J 2006 A group field theory for 3d quantum gravity coupled to a scalar field *Class. Quant. Grav.* **23** 6543–76 (*Preprint* gr-qc/0602010)
[6] Born M 1938 A suggestion for unifying quantum theory and relativity *Proc. R. Soc. Lond.* A **165** 291.
[7] Smolin L 2005 Loop quantum gravity and Planck-scale phenomenology *Lect. Notes Phys.* **669** 363.
[8] Girelli F, Livine E R and Oriti D 2010 4d deformed special relativity from group field theories *Phys. Rev.* D**81** 024015 (*Preprint* 0903.3475 [gr-qc])
[9] Majid S 2000 Meaning of noncommutative geometry and the Planck-scale quantum group *Lect. Notes Phys.* **541** 227 (*Preprint* hep-th/0006166)
[10] Chang L N, Lewis Z, Minic D and Takeuchi T 2011 On the minimal length uncertainty relation and the foundations of string theory *Adv. High Energy Phys.* **2011** 493514 (*Preprint* 1106.0068 [hep-th])
[11] Amelino-Camelia G 2002 Relativity in space-times with short-distance structure governed by an observer-independent (Planckian) length scale *Int. J. Mod. Phys.* D**11** 35 (*Preprint* gr-qc/0012051)

[12] Amelino-Camelia G 2001 Testable scenario for relativity with minimum-length *Phys. Lett. B***510** 255 (*Preprint* hep-th/0012238)

[13] Kowalski-Glikman J 2001 Observer-independent quanta of mass and length *Phys. Lett.* A**286** 391–4 (*Preprint* hep-th/0102098)

[14] Smolin L and Magueijo J 2003 Generalized Lorentz invariance with an invariant energy scale *Phys. Rev.* D**67** 044017 (*Preprint* gr-qc/0207085)

[15] Smolin L and Magueijo J 2004 Gravity's Rainbow *Class. Quant. Grav.* **21** 1725–36 (*Preprint* gr-qc/0305055)

[16] Kowalski-Glikman J 2005 Introduction to doubly-special relativity *Lect. Notes Phys.* **669** 131–59 (*Preprint* hep-th/0405273)

[17] Amelino-Camelia G 2010 Doubly-special relativity: Facts, myths and some key open issues *Symmetry* **2** 230–71 (*Preprint* 1003.3942 [gr-qc])

[18] Amelino-Camelia G, Ellis J, Mavromatos N E, Nanopoulos D V and Sarkar S 1998 Tests of quantum gravity from observations of γ-ray bursts *Nature* **393** 763 (*Preprint* astro-ph/9712103)

[19] Ellis J, Mavromatos N E and Nanopoulos D V 2009 Probing a Possible Vacuum Refractive Index with Gamma-Ray Telescopes *Phys. Lett.* B**674** 83–6 (*Preprint* 0901.4052 [astro-ph.HE])

[20] Amelino-Camelia G and Smolin L 2009 Prospects for constraining quantum gravity dispersion with near term observations *Phys. Rev.* D**80** 084017 (*Preprint* 0906.3731 [astro-ph.HE])

[21] Abdo A A et al. [Fermi LAT/GBM Collaborations] 2009 A limit on the variation of the speed of light arising from quantum gravity effects *Nature* **462** 331 (*Preprint* 0908.1832 [astro-ph.HE])

[22] Amelino-Camelia G, Arzano M, Gubitosi G and Magueijo J 2013 Dimensional reduction in the sky *Phys. Rev.* D**87** 123532 (*Preprint* 1305.3153 [gr-qc])

[23] Amelino-Camelia G, Arzano M, Gubitosi G and Magueijo J 2014 Planck-scale dimensional reduction without a preferred frame *Phys. Lett.* B**736** 317–20 (*Preprint* 1311.3135 [gr-qc])

[24] Amelino-Camelia G 2005 Introduction to Quantum-Gravity Phenomenology *Lect. Notes Phys.* **669** 59-100 (*Preprint* gr-qc/0412136)

[25] Amelino-Camelia G, Freidel L, Kowalski-Glikman J and Smolin L 2011 The principle of relative locality *Phys. Rev.* D**84** 084010 (*Preprint* 1101.0931 [hep-th])

[26] Amelino-Camelia G, Arzano M, Kowalski-Glikman J, Rosati G and Trevisan G 2012 Relative-locality distant observers and the phenomenology of momentum-space geometry *Class. Quant. Grav.* **29** 075007 (*Preprint* 1107.1724 [hep-th])

[27] Amelino-Camelia G, Arzano M, Bianco S and Buonocore R J 2013 The DSR-deformed relativistic symmetries and the relative locality of 3D quantum gravity *Class. Quant. Grav.* **30** 065012 (*Preprint* 1210.7834 [hep-th])

[28] Amelino-Camelia G, Matassa M, Mercati F and Rosati G 2011 Taming Nonlocality in Theories with Planck-Scale Deformed Lorentz Symmetry *Phys. Rev. Lett.* **106** 071301 (*Preprint* 1006.2126 [gr-qc])

[29] Freidel L and Smolin L 2011 Gamma ray burst delay times probe the geometry of momentum space *Preprint* 1103.5626 [hep-th]

[30] Amelino-Camelia G, Barcaroli L and Loret N 2012 Modeling transverse relative locality **Int. J. Theor. Phys. 51** 3359 ((*Preprint* 1107.3334 [hep-th])

[31] Batista E and Majid S 2003 Noncommutative geometry of angular momentum space $U(su(2))$ *J. Math. Phys.* **44** 107 (*Preprint* hep-th/0205128)

[32] Majid S and Ruegg H 1994 Bicrossproduct structure of kappa Poincare group and noncommutative geometry *Phys. Lett.* B**334** 348 (*Preprint* hep-th/9405107)

[33] Amelino-Camelia G, Astuti V and Rosati G 2013 Relative locality in a quantum spacetime and the pregeometry of κ-Minkowski *Eur. Phys. J.* C**73** 2521 (*Preprint* 1206.3805 [hep-th])

Journal of Physics: Conference Series **634** (2015) 012004 doi:10.1088/1742-6596/634/1/012004

Triproducts, nonassociative star products and geometry of R-flux string compactifications

Paolo Aschieri[1,a] **and Richard J. Szabo**[2,b]

[1]Dipartimento di Scienze e Innovazione Tecnologica
and INFN Torino, Gruppo collegato di Alessandria
Università del Piemonte Orientale
Viale T. Michel 11, 15121 Alessandria, Italy
[2]Department of Mathematics, Heriot-Watt University
Colin Maclaurin Building, Riccarton, Edinburgh EH14 4AS, U.K.
and Maxwell Institute for Mathematical Sciences, Edinburgh, U.K.
and The Higgs Centre for Theoretical Physics, Edinburgh, U.K.

E-mail: [a]`aschieri@unipmn.it`, and [b]`R.J.Szabo@hw.ac.uk`

Abstract.

We elucidate relations between different approaches to describing the nonassociative deformations of geometry that arise in non-geometric string theory. We demonstrate how to derive configuration space triproducts exactly from nonassociative phase space star products and extend the relationship in various directions. By foliating phase space with leaves of constant momentum we obtain families of Moyal–Weyl type deformations of triproducts, and we generalize them to new triproducts of differential forms and of tensor fields. We prove that nonassociativity disappears on-shell in all instances. We also extend our considerations to the differential geometry of nonassociative phase space, and study the induced deformations of configuration space diffeomorphisms. We further develop general prescriptions for deforming configuration space geometry from the nonassociative geometry of phase space, thus paving the way to a nonassociative theory of gravity in non-geometric flux compactifications of string theory.

1. Introduction and summary

Noncommutative geometry has long been believed to provide a framework for understanding the generalizations of classical spacetime geometry required to describe Planck scale quantum geometry, and ultimately quantum gravity. Yet its precise relation with other approaches to quantum gravity such as string theory have remained somewhat elusive. The surge of interest in noncommutative geometry from string theory originally came about in the *open* string sector, where the massless bosonic modes contain gauge and scalar fields: It was observed that D-brane worldvolumes acquire a noncommutative deformation in the background of a non-zero B-field [13, 31, 32], which is moreover nonassociative when the flux $H = \mathrm{d}B$ is non-vanishing [15]; the low-energy worldvolume effective field theory is then described by a noncommutative gauge theory. However, this realization does not really shed light on connections with the *closed* string sector whose massless bosonic modes contain gravitational degrees of freedom such as the metric g, the B-field, and the dilaton ϕ. Vanishing of the beta-functions for these fields (at one-loop) as required by conformal invariance of the worldsheet field theory yields equations of motion which can be derived from the standard bosonic closed string low-energy effective action

$$S = \int_M R(g) - \frac{1}{12}\,\mathrm{e}^{-\phi/3}\,H \wedge *_g H - \frac{1}{6}\,\mathrm{d}\phi \wedge *_g \mathrm{d}\phi\ . \tag{1.1}$$

It can be hoped that an equivalent noncommutative/nonassociative version of this effective field theory provides a suitable target space framework in which to address problems related to quantum gravity.

A precise connection between noncommutative geometry and the closed string sector has been found recently through non-geometric flux compactifications [21, 33]. The prototypical example involves a three-torus with constant three-form H-flux; T-dualizing along its cycles gives rise to constant geometric and non-geometric fluxes which can be depicted schematically through the T-duality chain

$$H_{abc} \xrightarrow{T_a} f^a{}_{bc} \xrightarrow{T_b} Q^{ab}{}_c \xrightarrow{T_c} R^{abc}\ . \tag{1.2}$$

While the H-flux and metric f-flux backgrounds can be described globally as Riemannian manifolds, the Q-flux background involves T-duality transformations as transition functions between local trivializations of its tangent bundle and gives rise to a T-fold, while the R-flux background is a purely non-geometric string vacuum as its metric and B-field are not even locally defined because they depend on the winding coordinates of the dual space. These non-geometric flux compactifications have been recently purported to have a global description in terms of noncommutative and nonassociative structures [24, 10, 12, 26] (see [28] for a recent review and further references).

Just as in the case of open strings, there are two ways in which one can see the appearence of nonassociativity of target space coordinates in the R-flux background: Either through canonical analysis of the closed string coordinates or by careful examination of worldsheet scattering amplitudes. In the formulation of [24], nonassociativity of the coordinates of *configuration space* arises as a failure of the Jacobi identity of the bracket of canonical variables on *phase space*. This model is described in Section 2, and it defines a twisted Poisson structure on phase space which can be quantized using Kontsevich formality to obtain an explicit nonassociative star product of functions for deformation quantization of phase space [26]. The nonassociative star product on phase space defines nonassociative tori in closed string theory. If we regard it as a substitute for canonical quantization as suggested by [6], then it may be used to formulate a consistent nonassociative version of quantum mechanics [27]. On the other hand, from a worldsheet perspective an exact analysis of conformal field theory correlation functions can be carried out to linear order in the H-flux, wherein the backreaction due to the curvature of spacetime can be neglected. This was performed by [12] where off-shell traces of nonassociativity were found and packaged into "triproducts" of fields directly on configuration space at linear order in R; at this order the elementary triproduct of three fields was shown by [26] to be reproduced by the associator of the phase space star product. In [12] an all orders exponential formula for arbitrary n-triproducts involving n fields was moreover conjectured, however its verification through a conformal field theory analysis on flat space is not possible.

In this contribution we fill a technical gap in these two approaches to the nonassociative geometry of R-flux compactifications by explicitly *deriving* the conjectural all orders configuration space n-triproducts of [12] from the phase space star products of [26]. As their origins are quite different, this provides a non-trivial confirmation of these triproducts and moreover of the validity of the phase space formulation as the fundamental theory of closed strings propagating in the non-geometric R-flux compactification, which is the perspective advocated by [26] and pursued in the present paper wherein phase space appears as the effective closed string target space. Physical applications of this perspective to scalar field theories can be found in [25].

For later reference, let us briefly review the phase space approach. The starting point of [26] is the Courant sigma-model based on the standard Courant algebroid $C = TM \oplus T^*M$ over the target space M with natural frame $(\psi_I) = (\partial_i, \mathrm{d}x^i)$; the structure maps consist of the fibre metric h_{IJ} with $h_{ij} = \langle \partial_i, \mathrm{d}x^j \rangle = \delta_i{}^j$, the anchor matrix $\rho(\psi_I) = P_I{}^i(x)\, \partial_i$,[1] and the three-form $T_{IJK}(x) = [\psi_I, \psi_J, \psi_K]_C$. This data can be used to construct a topological field theory on a three-dimensional M2-brane worldvolume Σ_3 with action given by

$$S_T = \int_{\Sigma_3} \phi_i \wedge \mathrm{d}X^i + \frac{1}{2}\, h_{IJ}\, \alpha^I \wedge \mathrm{d}\alpha^J - P_I{}^i(X)\, \phi_i \wedge \alpha^I + \frac{1}{6}\, T_{IJK}(X)\, \alpha^I \wedge \alpha^J \wedge \alpha^K \ , \qquad (1.3)$$

where $X : \Sigma_3 \to M$ is the M2-brane embedding, and $\alpha \in \Omega^1(\Sigma_3, X^*C)$ and $\phi \in \Omega^2(\Sigma_3, X^*T^*M)$ are auxilliary fields. By taking T to be any of the fluxes in (1.2), one can look at the effective dynamics of the membrane fields; putting in all of the fluxes simultaneously and independently of one another (e.g., by replacing M with its double manifold as in double field theory) could then give a precise formulation of the formal T-duality chain (1.2) in terms of (quantum) gauge symmetries of the Courant sigma-model. In the case of a purely geometric H-flux background it is shown in [26] that the M2-brane model reduces to the standard H-twisted Poisson sigma-model with target space M on the boundary $\Sigma_2 = \partial\Sigma_3$, which describes closed string propagation in the configuration space M with a non-constant B-field. On the other hand, in the case of a non-geometric constant R-flux the M2-brane model reduces instead to a generalized Poisson sigma-model on Σ_2 whose action can be expressed in linearized form by using auxilliary one-form fields η_I as

$$S_R = \int_{\Sigma_2} \eta_I \wedge \mathrm{d}X^I + \frac{1}{2}\, \Theta^{IJ}(X)\, \eta_I \wedge \eta_J \ , \qquad (1.4)$$

where $X = (X^i, P_i) : \Sigma_2 \to T^*M$ now embed closed strings into an effective target space which

[1] Implicit summation over repeated upper and lower indices is understood throughout.

coincides with the phase space of M, and

$$\Theta = \begin{pmatrix} R^{ijk}\,p_k & \delta^i{}_j \\ -\delta_i{}^j & 0 \end{pmatrix} \tag{1.5}$$

is a twisted Poisson bivector which describes a noncommutative and nonassociative phase space (see Section 2). This point of view that closed strings should propagate in phase space rather than configuration space is also pursued in a more general context by [17] wherein it is argued that the fundamental symmetry of string theory contains phase space diffeomorphisms in order to accomodate T-duality and other stringy features.

In Section 3 we shall in fact derive a generalization to families of Moyal–Weyl type deformations of the triproducts which are induced from a foliation of phase space by leaves of constant momentum and describe their physical significance; the leaf of zero momentum yields the triproducts of [12]. We further give an explicit proof that the integrated versions of these deformed triproducts coincide with those of the associative Moyal–Weyl star products, thus reproducing the physical expectations that nonassociativity is not visible in on-shell conformal field theory correlation functions. This is a remarkable cancellation that happens due to specific orderings of bracketing phase space star products. In Section 4 we generalize these results to new (deformed) triproducts of arbitrary differential forms via nonassociative deformations of phase space exterior products following the cochain twist deformation formalism proposed by [27] for the study of non-geometric fluxes. These algebraic techniques, originally developed to study quasi-Hopf algebra symmetries [16], are further extended to study the differential geometry of nonassociative phase space, i.e., the deformed Lie algebra of infinitesimal diffeomorphisms and their Lie derivative action on exterior forms and tensor fields. This geometry is then induced on each constant momentum leaf leading in particular to a deformed Lie algebra of infinitesimal diffeomorphisms on configuration space.

While the results of the present paper are mostly technical, they offer further insight into the nature of nonassociative geometry in non-geometric string theory in at least three ways. Firstly, we give a fairly general prescription for inducing configuration space triproducts from nonassociative star products on phase space and clarify in what sense these products obey the physical expectations of cyclicity of on-shell string scattering amplitudes, though a complete geometric framework for choosing appropriate section slices remains to be worked out. The different leaves correspond to a multitude of choices of Bopp shifts in phase space and are reminescent of the section conditions in double field theory (the weak and strong constraints). Secondly, our considerations based on cochain twist theory are general enough that they may be applicable beyond the flat space limit to the *curved* space triproducts conjectured by [11] for non-constant fluxes, and hence they may provide a precise link between the phase space formalism and the framework of double field theory; such a connection is explored by [17] in a related but more general context. Thirdly, our derivation and considerations of new triproducts involving arbitrary differential forms and infinitesimal diffeomorphisms on configuration space is a first step to understanding the construction of nonassociative deformations of gravity and their physical relevance in on-shell string theory.

2. Nonassociative geometry of R-space

2.1. Phase space formulation

We shall begin by describing the phase space model for the non-geometric R-flux background and discuss some ways in which it may be interpreted as a deformation of the geometry of configuration space. Let $M = \mathbb{R}^d$ be the decompactification limit of a d-torus, endowed with a constant three-form H-flux. As reviewed in Section 1, via T-duality this space is mapped to a non-geometric string background with a constant trivector R-flux $R = \frac{1}{3!} R^{ijk}\, \partial_i \wedge \partial_j \wedge \partial_k$, where $\partial_i = \frac{\partial}{\partial x^i}$ in local coordinates $x = (x^i) \in \mathbb{R}^d$. Explicit string and conformal field theory computations show that the string geometry acquires a noncommutative and nonassociative deformation for closed strings which wind and propagate in the non-geometric background [10, 24, 12, 14, 2]. An explicit realization of such a nonassociative deformation of the spacetime geometry is provided by the phase space description of the parabolic R-flux model on the cotangent bundle $\mathcal{M} = T^*M = M \times (\mathbb{R}^d)^*$ with local coordinates (x,p), where $p = (p_i) \in (\mathbb{R}^d)^*$. In this setting, the deformation is described by a commutator algebra for the local phase space coordinates given by

$$[x^i, x^j] = \frac{\mathrm{i}\, \ell_s^4}{3\hbar}\, R^{ijk}\, p_k \ , \qquad [x^i, p_j] = \mathrm{i}\, \hbar\, \delta^i{}_j \qquad \text{and} \qquad [p_i, p_j] = 0 \ , \tag{2.1}$$

which has a non-trivial Jacobiator

$$[x^i, x^j, x^k] = \ell_s^4\, R^{ijk} \ . \tag{2.2}$$

In the point particle limit $\ell_s = 0$ this is just the usual Lie algebra of ordinary quantum phase space.

As anticipated by [10, 24, 11], and proven in [27] directly from the phase space model, the R-flux background does not permit the notion of a point due to a minimal volume which enters an uncertainty relation for the position coordinates given by

$$\Delta x^i\, \Delta x^j\, \Delta x^k \geq \tfrac{1}{2}\, \ell_s^4\, R^{ijk} \ . \tag{2.3}$$

This lack of a notion of a point illustrates why the R-flux compactification is not even locally geometric; this is evident in the phase space model which, as a result of T-duality, requires both position coordinates and momenta in any local description. Hence it is not clear how to formulate a gravity theory, or any other field theory, on this nonassociative space. An approach based on fundamental loop space variables, rather than functions on M, was pursued in [29, 30]; the usage of such variables is also mentioned in [11]. This approach reflects the "non-locality" of the non-geometric R-space. It is natural in the lift of Type IIA string theory to M-theory in which the (closed string) boundary of an open M2-brane ending on an M5-brane in a constant C-field background gives rise to a noncommutative loop space algebra on the M5-brane worldvolume [8, 23], which corresponds to the noncommutativity and nonassociativity felt by a fundamental closed string in a constant H-flux; this perspective is utilized in the formulation by [26] of closed string propagation in the non-geometric R-flux background using M2-brane degrees of freedom, as reviewed in Section 1, and it connects open and closed string noncommutative geometry.

Another set of fundamental variables is obtained by considering the algebra $\mathsf{Diff}(M)$ of (formal) differential operators on M with typical elements of the form

$$\underline{f}(x) = f(x) + \sum_{k=1}^{\infty} f^{i_1\cdots i_k}(x)\, \partial_{i_1} \cdots \partial_{i_k} \ , \tag{2.4}$$

where $f, f^{i_1\cdots i_k} \in A := C^\infty(M)$ and $\underline{f}(x)g(x) = f(x)\, g(x) + \sum_k f^{i_1\cdots i_k}(x)\, \partial_{i_1} \cdots \partial_{i_k} g(x)$ for all $g \in A$. From the last two commutation relations in (2.1) we can identify $p_i = -\mathrm{i}\, \hbar\, \partial_i$ and the nonassociativity relations as relations involving differential operators

$$[x^i, x^j] = \tfrac{1}{3}\, \ell_s^4\, R^{ijk}\, \partial_k \ . \tag{2.5}$$

Such an interpretation was advocated by [19, 18] in the (associative) context of open strings on D-branes in non-constant B-field backgrounds.

A related interpretation of the twisted Poisson structure (2.1) on phase space $\mathcal{M} = T^*M$, as discussed in [26, Section 2.5], is that of a higher Poisson structure on the multivector field algebra $\mathsf{Vect}^\bullet(M) = C^\infty(M, \bigwedge^\bullet TM)$.

2.2. Deformations of configuration space geometry

In order to formulate a nonassociative theory of gravity on the original configuration space M, we note that there is a natural foliation of $\mathcal{M}$ over M defined by the (global) sections $s_{\bar p} : M \to \mathcal{M}$ with $s_{\bar p}(x) = (x, \bar p)$ for $\bar p \in (\mathbb{R}^d)^*$; this simply means that we can realise M as any leaf of constant momentum in phase space. We can therefore restrict functions on phase space $\underline{f} \in C^\infty(\mathcal{M})$ to $C^\infty(M)$ via the family of pullbacks $f_{\bar p} := s_{\bar p}^* \underline{f}$. Thus via pullback along the bundle projection $\pi : \mathcal{M} \to M$ we can transport functions on M into the cotangent bundle, perform the necessary nonassociative deformations on $\mathcal{M}$, and then pullback along $s_{\bar p}$ to obtain the desired nonassociative deformations of the geometry of M; schematically, we can depict this foliation by quantizations of the configuration manifold M via the commutative diagrams

$$\begin{array}{ccc} C^\infty(\mathcal{M}) & \xrightarrow{\ \mathscr{Q}\ } & \widehat{C^\infty(\mathcal{M})} \\ {\scriptstyle \pi^*}\uparrow & & \downarrow{\scriptstyle s_{\bar p}^*} \\ C^\infty(M) & \underset{\mathscr{Q}_{\bar p}}{\dashrightarrow} & \widehat{C^\infty(M)} \end{array} \tag{2.6}$$

with $s_{\bar p}^* \circ \pi^* = (\pi \circ s_{\bar p})^* = \mathrm{id}$, where $\mathscr{Q}$ is the nonassociative quantization of $\mathcal{M}$ whose foundations are developed by [26, 27], while $\mathscr{Q}_{\bar p}$ is the desired nonassociative quantization of M induced by the diagram. We can use this perspective to obtain nonassociative field theories on M via this systematic implementation of the nonassociative geometry of $\mathcal{M}$.

This perspective also suggests a way in which to obtain richer deformations of the geometry of configuration space from the nonassociative quantization of phase space. By pullbacks along π^* we consider fields $\underline{f} \in C^\infty(\mathcal{M})$ that satisfy the *section constraints*

$$\tilde\partial^i\, \underline{f} = 0 \qquad \text{for} \quad i = 1, \dots, d \ , \tag{2.7}$$

where $\tilde\partial^i := \frac{\partial}{\partial p_i}$. There is a natural invariant metric on phase space $\mathcal{M} = T^*M$ which in a local frame is defined by

$$\gamma = \mathrm{d}x^i \otimes \mathrm{d}p_i + \mathrm{d}p_i \otimes \mathrm{d}x^i \ . \tag{2.8}$$

It defines an $O(d,d)$-structure on $\mathcal{M}$, i.e., a reduction of the structure group $GL(2d, \mathbb{R})$ of the tangent bundle $T\mathcal{M}$ to the subgroup $O(d,d)$; this is the symmetry group underlying quantum mechanical Born reciprocity and its relation to T-duality is explained by [17]. Drawing now on the evident similarities with double field theory (see [1, 20] for reviews and [34] for a rigorous

Journal of Physics: Conference Series **634** (2015) 012004 doi:10.1088/1742-6596/634/1/012004

mathematical treatment), a natural weakening of the section condition (2.7) involves constraints on phase space functions $\underline{f}$ which remove their momentum dependence locally only up to an $O(d,d)$ transformation. For example, if θ is any constant nondegenerate bivector on M, then this constraint is solved by any phase space function $\underline{f}$ which depends only on the non-local Bopp shifts $x+\theta\cdot p$; this d-dimensional slice of the $2d$-dimensional phase space geometry can be rotated to the constant momentum leaves above via an $O(d,d)$ transformation. In particular, the tangent bundle on phase space decomposes as $T\mathcal{M}\cong L\oplus L^*$ where L is the tangent bundle on the leaves of the foliation and L^* is its dual bundle with respect to the orthogonal complement in the metric (2.8). The section constraints on functions $\underline{f}\in C^\infty(\mathcal{M})$ and more generally on covariant tensor fields $\underline{T}\in C^\infty(\mathcal{M},\bigotimes^\bullet T^*\mathcal{M})$ now read as

$$Z(\underline{f}\,)=0 \qquad \text{and} \qquad \imath_Z\,\underline{T}=0=\mathcal{L}_Z\,\underline{T} \tag{2.9}$$

for all sections $Z\in C^\infty(\mathcal{M},L^*)$, where $\imath$ denotes contraction and $\mathcal{L}$ is the Lie derivative. This means that the set of admissible fields is constrained to *foliated* tensor fields with respect to the distribution L^*. For the foliation defined by $s_{\bar{p}}$, one has $L=TM$ and the section constraints can be imposed by taking $Z=\tilde{\partial}^i$ for each $i=1,\dots,d$. A similar perspective on closed string target spaces is addressed in [17] where connections with double field theory are also explored.

These more general foliations of phase space could lead to much richer classes of deformations of M, because the nonassociative quantizations $\mathscr{Q}$ are *not* $O(d,d)$-invariant, as is evident from the defining relations (2.1). Moreover, quantization of the section conditions (2.9) themselves could lead to interesting deformations of the foliated tensor fields on the d-dimensional slices. It would be interesting to repeat the analysis of this paper for such more general section slices. Among other things, this could help elucidate possible relationships between the noncommutative gerbe structure [3] on phase space underlying the nonassociative deformations [26] and the abelian gerbe structures underlying the generalised manifolds in double geometry [9, 22]. It should also help in understanding how to lift geometric objects from M to its cotangent bundle $\mathcal{M}=T^*M$ in a way suited to describe nonassociative deformations of the geometry and of gravity directly on the configuration manifold M.

3. Triproducts from phase space star products

3.1. Families of n-triproducts

We shall now derive triproduct formulas in the context of Section 2 which include those of [12] as special cases. Our starting point is the nonassociative star product $\star$ on phase space derived by [26] (see also [6]) which quantizes the commutation relations (2.1). Here we shall utilize the expression for this star product as a twisted convolution product (see Section 4.3 for a different derivation). For phase space functions $\underline{f},\underline{g}\in C^\infty(\mathcal{M})$, the integral formula derived in [27, eq. (3.38)] adapted to the normalizations in (2.1) reads as

$$\begin{aligned}(\underline{f}\star\underline{g})(x,p) \;=\; &\frac{1}{(\pi\,\hbar)^{2d}}\iint_M \mathrm{d}^dz\;\mathrm{d}^dz'\;\iint_{M^*}\mathrm{d}^dk\;\mathrm{d}^dk'\;\underline{f}(x+z,p+k)\,\underline{g}(x+z',p+k')\\ &\times\;\mathrm{e}^{-\frac{2\,\mathrm{i}}{\hbar}\,(k\cdot z'-k'\cdot z)}\;\mathrm{e}^{-\frac{2\,\mathrm{i}\,\ell_s^4}{3\hbar^3}\,R(k,k',p)}\;,\end{aligned} \tag{3.1}$$

where $k'\cdot z:=k_i'\,z^i$ and $R(k,k',p):=R^{ijl}\,k_i\,k_j'\,p_l$. Restricted to functions $f,g\in C^\infty(M)$ by the pullback along $\pi:\mathcal{M}\to M$, after rescaling, this product reads as

$$(\pi^*f\star\pi^*g)(x,p) \;=\; \frac{1}{(2\pi)^{2d}}\iint_M \mathrm{d}^dz\;\mathrm{d}^dz'\;f(x+z)\,g(x+z')$$

$$\times \iint_{M^*} \mathrm{d}^d k \; \mathrm{d}^d k' \;\, \mathrm{e}^{\,-\mathrm{i}\,(k\cdot z' - k'\cdot z)} \;\, \mathrm{e}^{\,-\frac{\mathrm{i}\,\ell_s^4}{6\hbar}\,R(k,k',p)} \;. \tag{3.2}$$

In general, the star product of two fields on M is a field on $\mathcal{M}$, i.e., a differential operator on M. Thus the star product $\star$ does not close in the algebra $A = C^\infty(M)$. We can define a family of products of fields on M by pulling this product back under the local sections $s_{\bar p} : M \to \mathcal{M}$ that foliate $\mathcal{M}$ with leaves isomorphic to M. This defines a family of 2-products $\mu^{(2)}_{\bar p} : A \otimes A \to A$ given by

$$\begin{aligned}
\mu^{(2)}_{\bar p}(f\otimes g)(x) \;&:=\; s^*_{\bar p}\,(\pi^* f \star \pi^* g)(x,p) \\
&= \frac{1}{(2\pi)^{2d}} \iint_M \mathrm{d}^d z \; \mathrm{d}^d z' \;\, f(x+z)\, g(x+z'\,) \\
&\qquad\qquad \times \iint_{M^*} \mathrm{d}^d k \; \mathrm{d}^d k' \;\, \mathrm{e}^{\,-\mathrm{i}\,(k\cdot z'-k'\cdot z)} \;\, \mathrm{e}^{\,-\frac{\mathrm{i}\,\ell_s^4}{6\hbar}\,\theta_{\bar p}(k,k'\,)} \\
&= \frac{1}{(2\pi)^{2d}} \iint_{M^*} \mathrm{d}^d k \; \mathrm{d}^d k' \;\, \mathrm{e}^{\,\mathrm{i}\,(k+k'\,)\cdot x} \;\, \mathrm{e}^{\,\frac{\mathrm{i}\,\ell_s^4}{6\hbar}\,\theta_{\bar p}(k,k'\,)} \\
&\qquad\qquad \times \iint_M \mathrm{d}^d z \; \mathrm{d}^d z' \;\, f(z)\, g(z'\,) \;\, \mathrm{e}^{\,-\mathrm{i}\,(k\cdot z'+k'\cdot z)} \\
&= \int_{M^*} \mathrm{d}^d k \; \int_{M^*} \mathrm{d}^d k' \;\, \hat f(k'\,)\, \hat g(k)\, \mathrm{e}^{\,\mathrm{i}\,(k+k'\,)\cdot x} \;\, \mathrm{e}^{\,\frac{\mathrm{i}\,\ell_s^4}{6\hbar}\,\theta_{\bar p}(k,k'\,)} \\
&= \int_{M^*} \mathrm{d}^d k \; \int_{M^*} \mathrm{d}^d k' \;\, \hat f(k)\, \hat g(k'\,)\, \mathrm{e}^{\,\mathrm{i}\,(k+k'\,)\cdot x} \;\, \mathrm{e}^{\,-\frac{\mathrm{i}\,\ell_s^4}{6\hbar}\,\theta_{\bar p}(k,k'\,)} \;,
\end{aligned} \tag{3.3}$$

where

$$\theta_{\bar p} = \tfrac12\,\theta^{ij}_{\bar p}\;\partial_i\wedge\partial_j := \tfrac12\,R^{ijk}\;\bar p_k\;\partial_i\wedge\partial_j \tag{3.4}$$

are constant bivectors on M along the leaves of constant momentum in phase space, and

$$\hat f(k) = \frac{1}{(2\pi)^d} \int_M \mathrm{d}^d x \;\, f(x)\;\, \mathrm{e}^{\,-\mathrm{i}\,k\cdot x} \tag{3.5}$$

is the Fourier transform of the function $f \in C^\infty(M)$. This last expression is simply the momentum space representation of the associative Moyal–Weyl star product $\star_{\bar p}$ determined by the bivector $-\frac{\ell_s^4}{3\hbar}\,\theta_{\bar p}$, and we thus find an expression for the 2-products in terms of a bidifferential operator

$$\mu^{(2)}_{\bar p}(f\otimes g) = f \star_{\bar p} g := \mu\Big(\exp\big(\tfrac{\mathrm{i}\,\ell_s^4}{6\hbar}\,\theta^{ij}_{\bar p}\;\partial_i\otimes\partial_j\big)(f\otimes g)\Big) \tag{3.6}$$

where $\mu(f\otimes g) = f\,g$ is the pointwise multiplication of functions in $A = C^\infty(M)$.

We shall now generalize this result to n-products $\mu^{(n)}_{\bar p} : A^{\otimes n} \to A$ of functions $f_1,\dots,f_n \in A$ for $n \geq 3$, which we define in an analogous way. However, we must keep in mind that the phase space star product $\star$ is nonassociative, so we have to keep track of the order in which we group binary products of functions on $\mathcal{M}$; this is even true after integration over $\mathcal{M}$ [27]. We choose the ordering

$$\mu^{(n)}_{\bar p}(f_1\otimes\cdots\otimes f_n) := s^*_{\bar p}\big[\big((\cdots(\pi^* f_1 \star \pi^* f_2) \star \pi^* f_3) \star\cdots\big) \star \pi^* f_n\big] \;, \tag{3.7}$$

and discuss the significance of the other orderings below. We shall compare this n-product to iterations of the Moyal–Weyl star products given by

$$\star_{\bar p}(f_1\otimes\cdots\otimes f_n) := f_1\star_{\bar p}\cdots\star_{\bar p} f_n = \mu\Big[\exp\Big(\frac{\mathrm{i}\,\ell_s^4}{6\hbar}\sum_{1\le a<b\le n}\theta^{ij}_{\bar p}\,\partial_i^a\,\partial_j^b\Big)(f_1\otimes\cdots\otimes f_n)\Big]\ , \tag{3.8}$$

where $\partial_i^a = (\mathrm{id}\otimes\cdots\otimes\partial_i\otimes\cdots\otimes\mathrm{id})$, $a=1,\dots,n$ denotes the derivative ∂_i acting on the a-th factor of the tensor product $f_1\otimes\cdots\otimes f_n$; note that no bracketings need be specified here due to associativity of the products $\star_{\bar p}$ for all $\bar p$.

Proposition 3.9.

$$\mu^{(n)}_{\bar p}(f_1\otimes\cdots\otimes f_n) = \star_{\bar p}\Big[\exp\Big(\frac{\ell_s^4}{12}\sum_{1\le a<b<c\le n}R^{ijk}\,\partial_i^a\,\partial_j^b\,\partial_k^c\Big)(f_1\otimes\cdots\otimes f_n)\Big]\ .$$

Proof. By iterating the integral formula (3.1), it is straightforward to derive the integral representation

$$\begin{aligned} s^*_{\bar p}\big[((\cdots(\pi^*f_1\star\pi^*f_2)\star\pi^*f_3)\star\cdots)\star\pi^*f_n\big](x) \\ = \frac{1}{(\pi\,\hbar)^{2(n-1)d}}\prod_{a=1}^{n-1}\iint_M \mathrm{d}^dz_a\,\mathrm{d}^dz'_a\iint_{M^*}\mathrm{d}^dk_a\,\mathrm{d}^dk'_a\ \mathrm{e}^{-\frac{2\,\mathrm{i}}{\hbar}\,(k_a\cdot z'_a-k'_a\cdot z_a)} \\ \times\prod_{b=1}^{n} f_b(x+z_1+\cdots+z_{n-b}+z'_{n-b+1}) \\ \times\ \exp\Big(-\frac{2\,\mathrm{i}\,\ell_s^4}{3\hbar^3}\sum_{c=1}^{n-1}R(k_c,k'_c,p+k_1+\cdots+k_{c-1})\Big) \end{aligned} \tag{3.10}$$

with the conventions $z_0=0, z'_n=0, k_0=0$. After rescaling we thus get

$$\begin{aligned} \mu^{(n)}_{\bar p}(f_1\otimes\cdots\otimes f_n)(x) &= \frac{1}{(2\pi)^{2(n-1)d}}\prod_{a=1}^{n-1}\iint_M \mathrm{d}^dz_a\,\mathrm{d}^dz'_a\iint_{M^*}\mathrm{d}^dk_a\,\mathrm{d}^dk'_a\ \mathrm{e}^{-\mathrm{i}\,(k_a\cdot z'_a-k'_a\cdot z_a)} \\ &\qquad\times\prod_{b=1}^{n} f_b(x+z_1+\cdots z_{n-b}+z'_{n-b+1}) \\ &\qquad\times\ \exp\Big(-\frac{\mathrm{i}\,\ell_s^4}{12}\sum_{c=1}^{n-1}R(k_c,k'_c,k_1+\cdots+k_{c-1})\Big) \\ &\qquad\times\ \exp\Big(-\frac{\mathrm{i}\,\ell_s^4}{6\hbar}\sum_{c=1}^{n-1}\theta_{\bar p}(k_c,k'_c)\Big) \\ &= \frac{1}{(2\pi)^{2(n-1)d}}\prod_{a=1}^{n-1}\iint_M \mathrm{d}^dz_a\,\mathrm{d}^dz'_a\iint_{M^*}\mathrm{d}^dk_a\,\mathrm{d}^dk'_a\ \mathrm{e}^{-\mathrm{i}\,(k_a\cdot z'_a-k'_a\cdot z_a)} \\ &\qquad\times\ \mathrm{e}^{\,\mathrm{i}\,(k_1+\cdots+k_{n-1}-k'_{n-1})\cdot x}\ \mathrm{e}^{-\mathrm{i}\,k'_{n-1}\cdot(z_1+\cdots+z_{n-2})} \\ &\qquad\times\mathrm{e}^{\,\mathrm{i}\,k_a\cdot(z_1+\cdots+z_{a-1})}\prod_{b=2}^{n} f_b(z'_{n-b+1})\ f_1(z_{n-1}) \end{aligned}$$

$$\times \exp\Big(-\frac{\mathrm{i}\,\ell_s^4}{12}\sum_{c=1}^{n-1} R(k_c,k_c',k_1+\cdots+k_{c-1})\Big)$$
$$\times \exp\Big(-\frac{\mathrm{i}\,\ell_s^4}{6\hbar}\sum_{c=1}^{n-1} \theta_{\bar p}(k_c,k_c')\Big) \qquad (3.11)$$

where we relabelled $z'_{n-b+1} \to z'_{n-b+1} - (x + z_1 + \cdots + z_{n-b})$ for $2 \le b \le n$, and $z_{n-1} \to z_{n-1} - (x + z_1 + \cdots + z_{n-2})$. Integrating over z_a for $a = 1,\dots,n-2$ gives delta-function constraints

$$k_a' = k_{n-1}' - (k_{a+1} + \cdots + k_{n-1}) \qquad \text{for} \quad a = 1,\dots,n-2\ . \qquad (3.12)$$

Integrating over z'_a for $a = 1,\dots,n-1$ and over z_{n-1}, and after relabelling $k_1 \to k_n$, $k'_{n-1} \to -k_1$ and $k'_{n-b+1} \to k_b$ for $b = 2,\dots,n$, we then get

$$\begin{aligned}\mu_{\bar p}^{(n)}(f_1\otimes\cdots\otimes f_n)(x) &= \prod_{a=1}^{n}\int_{M^*} \mathrm{d}^d k_a\ \hat f_a(k_a)\ \mathrm{e}^{\,\mathrm{i}\,(k_1+\cdots+k_n)\cdot x} \\ &\qquad \times \exp\Big(\frac{\mathrm{i}\,\ell_s^4}{12}\sum_{c=1}^{n-1} R(k_{n-c+1},k_1+\cdots+k_{n-c},k_{n-c+2}+\cdots+k_n)\Big) \\ &\qquad \times \exp\Big(\frac{\mathrm{i}\,\ell_s^4}{6\hbar}\sum_{c=1}^{n-1} \theta_{\bar p}(k_{n-c+1},k_1+\cdots+k_{n-c})\Big) \\ &= \prod_{a=1}^{n}\int_{M^*} \mathrm{d}^d k_a\ \hat f_a(k_a)\ \mathrm{e}^{\,\mathrm{i}\,(k_1+\cdots+k_n)\cdot x} \\ &\qquad \times \exp\Big(-\frac{\mathrm{i}\,\ell_s^4}{12}\sum_{c=1}^{n-1}\sum_{a=1}^{n-c}\sum_{b=n-c+2}^{n} R(k_a,k_{n-c+1},k_b)\Big) \\ &\qquad \times \exp\Big(-\frac{\mathrm{i}\,\ell_s^4}{6\hbar}\sum_{c=1}^{n-1}\sum_{a=1}^{n-c} \theta_{\bar p}(k_a,k_{n-c+1})\Big) \end{aligned} \qquad (3.13)$$

where we used multilinearity and antisymmetry of the trivector and bivector terms. Rewriting the summations we finally arrive at

$$\begin{aligned}\mu_{\bar p}^{(n)}(f_1\otimes\cdots\otimes f_n)(x) &= \prod_{a=1}^{n}\int_{M^*} \mathrm{d}^d k_a\ \hat f_a(k_a)\ \mathrm{e}^{\,\mathrm{i}\,(k_1+\cdots+k_n)\cdot x} \\ &\qquad \times \exp\Big(-\frac{\mathrm{i}\,\ell_s^4}{12}\sum_{1\le a<e<b\le n} R(k_a,k_e,k_b)\Big) \\ &\qquad \times \exp\Big(-\frac{\mathrm{i}\,\ell_s^4}{6\hbar}\sum_{1\le a<b\le n} \theta_{\bar p}(k_a,k_b)\Big) \end{aligned} \qquad (3.14)$$

which is the asserted result written using the momentum space representation for the $\star_{\bar p}$ product and the R-flux differential operators. □

Besides the fundamental 2-products $\mu_{\bar p}^{(2)}$, which we have seen coincide with the Moyal–Weyl star products $\star_{\bar p}$, the basic triproducts are given by

$$\mu_{\bar p}^{(3)}(f\otimes g\otimes h) = \star_{\bar p}\Big(\exp\big(\tfrac{\ell_s^4}{12}\,R^{ijk}\,\partial_i\otimes\partial_j\otimes\partial_k\big)(f\otimes g\otimes h)\Big)\ . \qquad (3.15)$$

The coordinate space commutator in (2.1) is then reproduced by the commutator bracket for $\mu_{\bar{p}}^{(2)}$, while the 3-bracket (2.2) is reproduced by $[x^i, x^j, x^k]_{\bar{p}}$, where

$$[f_1, f_2, f_3]_{\bar{p}} := \sum_{\sigma\in S_3} (-1)^{|\sigma|}\, \mu_{\bar{p}}^{(3)}(f_{\sigma(1)} \otimes f_{\sigma(2)} \otimes f_{\sigma(3)}) \tag{3.16}$$

with S_3 the group of permutations of the set $\{1, 2, 3\}$. The general n-triproducts satisfy the reduction properties

$$\mu_{\bar{p}}^{(n)}(f_1 \otimes \cdots \otimes (f_i\!=\!1) \otimes \cdots \otimes f_n) \;=\; \mu_{\bar{p}}^{(n-1)}(f_1 \otimes \cdots \otimes \widehat{f_i} \otimes \cdots \otimes f_n) \tag{3.17}$$

where $\widehat{f_i}$ denotes omission of f_i, $i = 1, \ldots, n$; these reductions consistently yield

$$\mu_{\bar{p}}^{(3)}(f \otimes g \otimes 1) = \mu_{\bar{p}}^{(3)}(f \otimes 1 \otimes g) = \mu_{\bar{p}}^{(3)}(1 \otimes f \otimes g) = \mu_{\bar{p}}^{(2)}(f \otimes g) = f \star_{\bar{p}} g\ . \tag{3.18}$$

However, for $n > 2$ the n-triproducts $\mu_{\bar{p}}^{(n)}$ cannot be defined by iteration from m-triproducts with $m < n$. This is in contrast to the precursor definition (3.7) in terms of phase space star products, and also to the n-star products which are defined in (3.8) by iteration of the Moyal–Weyl star products.

One of the most distinctive features of the n-triproducts is their trivialization on-shell, i.e., after integration over M.

Proposition 3.19. $\displaystyle\int_M \mathrm{d}^d x\ \mu_{\bar{p}}^{(n)}(f_1 \otimes \cdots \otimes f_n) = \int_M \mathrm{d}^d x\ f_1 \star_{\bar{p}} \cdots \star_{\bar{p}} f_n\ .$

Proof. Using the momentum space integral formula (3.14) for the n-triproduct, we have

$$\begin{aligned}\int_M \mathrm{d}^d x\ \mu_{\bar{p}}^{(n)}(f_1 \otimes \cdots \otimes f_n) \;&=\; (2\pi)^d\, \prod_{a=1}^n\, \int_{M^*} \mathrm{d}^d k_a\ \tilde{f}_a(k_a)\ \delta(k_1 + \cdots + k_n) \\ &\qquad \times\ \exp\Big(-\frac{\mathrm{i}\,\ell_s^4}{12} \sum_{1\le a<b<c\le n} R(k_a, k_b, k_c)\Big) \\ &\qquad \times\ \exp\Big(-\frac{\mathrm{i}\,\ell_s^4}{6\hbar} \sum_{1\le a<b\le n} \theta_{\bar{p}}(k_a, k_b)\Big)\ ,\end{aligned} \tag{3.20}$$

where the delta-function enforcing momentum conservation arises by translation-invariance on M of the definition (3.7). Using multilinearity of the trivector R we have

$$\sum_{1\le a<b<c\le n} R(k_a, k_b, k_c) = \sum_{b=2}^{n-1}\, \sum_{a=1}^{b-1}\, \sum_{c=b+1}^{n} R(k_a, k_b, k_c) = \sum_{b=2}^{n-1} R\Big(\sum_{a=1}^{b-1} k_a\,,\ k_b\,,\ \sum_{c=b+1}^{n} k_c\Big)\ , \tag{3.21}$$

and each term in the sum over b vanishes by antisymmetry of R after imposing momentum conservation $\sum_{a=1}^n k_a = 0$. □

From Proposition 3.19 it follows that the leaf of zero momentum is singled out by the feature that on-shell there are no signs of noncommutativity or nonassociativity. On the special slice $\bar{p} = 0$ the 2-product

$$\mu_0^{(2)}(f \otimes g) = f\, g \tag{3.22}$$

is the ordinary multiplication of functions, while for $n \geq 3$ the expression for the n-triproduct from Proposition 3.9 becomes

$$\mu_0^{(n)}(f_1 \otimes \cdots \otimes f_n) = \mu\Big[\exp\Big(\frac{\ell_s^4}{12} \sum_{1\leq a<b<c\leq n} R^{ijk}\,\partial_i^a\,\partial_j^b\,\partial_k^c\Big)(f_1 \otimes \cdots \otimes f_n)\Big] \ . \tag{3.23}$$

These products coincide with the sequence of n-triproducts for $n \geq 2$ that was proposed by [12] from an analysis of off-shell closed string tachyon amplitudes in the toroidal flux model to linear order in the flux components R^{ijk}. By Proposition 3.19 they obey the on-shell condition

$$\int_M \mathrm{d}^d x\ \mu_0^{(n)}(f_1 \otimes \cdots \otimes f_n) = \int_M \mathrm{d}^d x\ f_1 \cdots f_n \ , \tag{3.24}$$

and hence all traces of nonassociativity disappear in closed string scattering amplitudes. Here we have reproduced this conjectural triproduct to all orders in R^{ijk} from the all orders phase space star product derived by [26]. The correspondence between the phase space star product restricted to functions on M and the triproducts of [12] was already noted by [26] for $n = 2, 3$ (see also [6]); here we have extended and generalised the correspondence to all $n > 3$.

The triproducts for $\bar{p} \neq 0$ lead to on-shell correlation functions which are cyclically invariant, by cyclicity of the Moyal–Weyl products $\star_{\bar{p}}$ for all $\bar{p}$. In this case the scattering amplitudes resemble those of *open* strings on a D-brane with two-form B-field inverse to the bivector $\theta_{\bar{p}}$. The appearence of the more general triproducts $\mu_{\bar{p}}^{(n)}$ is natural from the perspective of the open/closed string duality described in [26, Section 2.4], which is a crucial ingredient in the derivation of the nonassociative phase space star product from closed string correlation functions; in fact, in [24, Appendix] it is argued that closed string momentum and winding modes define a certain notion of D-brane in closed string theory. Moreover, they are natural in light of the momentum space noncommutative gerbe structure of the nonassociative phase space description of the parabolic R-flux background and the closed string Seiberg–Witten maps relating associative and nonassociative theories, as described in [26, Section 3.4]. Hence in the following we work with the general leaves of constant momentum in order to retain as much of the (precursor) phase space nonassociative geometry on M as possible in our analysis, with the understanding that bonafide closed string scattering amplitudes require setting $\bar{p} = 0$ at the end of the day. We can interpret this distinction in the following way: If we regard the nonassociative field theory constructed on phase space as the fundamental field theory of closed strings in the non-geometric R-flux frame (as was done in [26] and as we do throughout in the present paper), with the leaves of constant momentum being the on-shell physical sectors, then integrating over momenta traces out extra degrees of freedom leading to violation of associativity. From this perspective localizing on the $\bar{p} = 0$ leaf corresponds to working on a closed string vacuum as a lowest order approximation to the theory, while higher (M2-brane) excitations involve fluctuations out of the $\bar{p} = 0$ leaf and probe the whole structure of the full nonassociative field theory.

3.2. Association relations

In our definition (3.7) we used a particular bracketing for the star product of n functions on phase space. As this star product is nonassociative, it is natural to ask what happens when one chooses different orderings. It was shown by [27] that the different choices of associating the functions is controlled by an associator, which can be described by a tridifferential operator

$$\Phi = \exp\ \tfrac{\ell_s^4}{6}\, R^{ijk}\,\partial_i \otimes \partial_j \otimes \partial_k\big) \ . \tag{3.25}$$

Journal of Physics: Conference Series **634** (2015) 012004 doi:10.1088/1742-6596/634/1/012004

The associator relates the two different products of three functions on phase space $(\underline{f}\star\underline{g})\star\underline{h}$ and $\underline{f}\star(\underline{g}\star\underline{h})$. If we introduce the notation

$$\Phi(\underline{f}\otimes\underline{g}\otimes\underline{h}) =: \underline{f}^{\phi}\otimes\underline{g}^{\phi}\otimes\underline{h}^{\phi} \tag{3.26}$$

where summation over the index ϕ is understood, then the relation is

$$(\underline{f}\star\underline{g})\star\underline{h} = \underline{f}^{\phi}\star(\underline{g}^{\phi}\star\underline{h}^{\phi}) \ . \tag{3.27}$$

We can write this relation more algebraically as

$$\mu_\star\circ(\mu_\star\otimes\mathrm{id}) = \mu_\star\circ(\mathrm{id}\otimes\mu_\star)\circ\Phi \ , \tag{3.28}$$

where $\mu_\star(\underline{f}\otimes\underline{g}) := \underline{f}\star\underline{g}$ with the linear maps

$$\mu_\star\circ(\mu_\star\otimes\mathrm{id}) : (\mathcal{A}\otimes\mathcal{A})\otimes\mathcal{A}\longrightarrow\mathcal{A} \qquad \text{and} \qquad \mu_\star\circ(\mathrm{id}\otimes\mu_\star) : \mathcal{A}\otimes(\mathcal{A}\otimes\mathcal{A})\longrightarrow\mathcal{A} \ . \tag{3.29}$$

Here we regard the associator as a linear map

$$\Phi : (\mathcal{A}\otimes\mathcal{A})\otimes\mathcal{A}\longrightarrow\mathcal{A}\otimes(\mathcal{A}\otimes\mathcal{A}) \qquad \text{with} \quad (\underline{f}\otimes\underline{g})\otimes\underline{h}\stackrel{\Phi}{\longmapsto}\underline{f}^{\phi}\otimes(\underline{g}^{\phi}\otimes\underline{h}^{\phi}) \ , \tag{3.30}$$

where the parentheses emphasize that on the left-hand side the product $\mu_\star\circ(\mu_\star\otimes\mathrm{id})$ naturally acts while on the right-hand side the product $\mu_\star\circ(\mathrm{id}\otimes\mu_\star)$ acts. Applying these natural products to (3.30) yields the identity (3.27).

The operator Φ is a 3-cocycle in the Hopf algebra of translations and Bopp shifts in phase space $\mathcal{M}$ that we will discuss in Section 4; this means that it obeys the pentagon relations which states that the two possible ways of reordering the brackets from left to right $((\mathcal{A}\otimes\mathcal{A})\otimes\mathcal{A})\otimes\mathcal{A}\to\mathcal{A}\otimes(\mathcal{A}\otimes(\mathcal{A}\otimes\mathcal{A}))$ are equivalent, i.e., the diagram

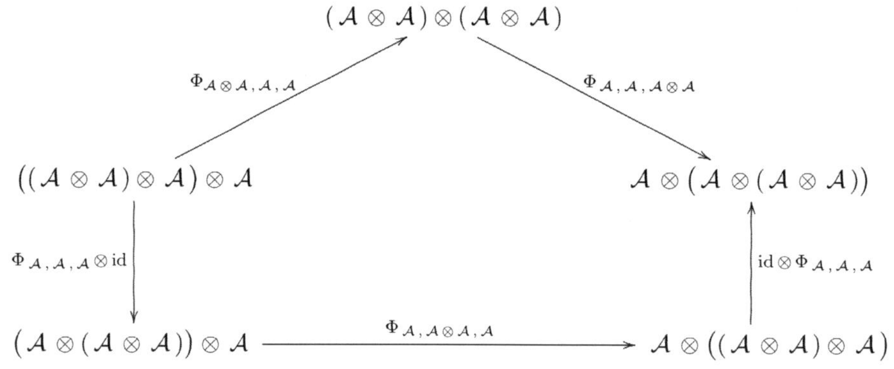

(3.31)

commutes. This means that we can rewrite the products $((\underline{f}\star\underline{g})\star\underline{h})\star\underline{k}$ as linear combinations of products of the kind $\underline{f}'\star(\underline{g}'\star(\underline{h}'\star\underline{k}'))$. There are *a priori* two different linear combinations, respectively obtained by following the upper and lower paths in the diagram (3.31). Commutativity of the diagram asserts that these two paths are equivalent. The action of the differential operator $\Phi_{\mathcal{A}\otimes\mathcal{A},\mathcal{A},\mathcal{A}}$ on $(\underline{f}\otimes\underline{g})\otimes\underline{h}\otimes\underline{k}$ is inherited from the Leibniz rule $\partial_i(\underline{f}\otimes\underline{g}) = \partial_i\underline{f}\otimes\underline{g}+\underline{f}\otimes\partial_i\underline{g}$ and it reads as

$$\Phi_{\mathcal{A}\otimes\mathcal{A},\mathcal{A},\mathcal{A}}\big((\underline{f}\otimes\underline{g})\otimes\underline{h}\otimes\underline{k}\big) = \exp\big(\tfrac{\ell_s^4}{6}\,R^{ijk}\,(\partial_i^1\,\partial_j^3\,\partial_k^4+\partial_i^2\,\partial_j^3\,\partial_k^4)\big)(\underline{f}\otimes\underline{g}\otimes\underline{h}\otimes\underline{k}) \ . \tag{3.32}$$

If we apply the product $\mu_\star\circ(\mu_\star\otimes\mu_\star)$ to this expression we obtain

$$\mu_\star\circ(\mu_\star\otimes\mu_\star)\circ\Phi_{\mathcal{A}\otimes\mathcal{A},\mathcal{A},\mathcal{A}}\big((\,\underline{f}\otimes\underline{g}\,)\otimes\underline{h}\otimes\underline{k}\,\big)=\big((\,\underline{f}\star\underline{g}\,)\star\underline{h}\,\big)\star\underline{k}\ , \tag{3.33}$$

as is read off from the upper arrow in (3.31). The differential operators $\Phi_{\mathcal{A},\mathcal{A},\mathcal{A}\otimes\mathcal{A}}$ and $\Phi_{\mathcal{A},\mathcal{A}\otimes\mathcal{A},\mathcal{A}}$ are similarly defined.

The associator Φ also relates the two inequivalent triproducts of functions on M given by $s_{\bar p}^*\big[(\pi^*f\star\pi^*g)\star\pi^*h\big]$ and $s_{\bar p}\big[\pi^*f\star(\pi^*g\star\pi^*h)\big]$. In order to present an explicit expression for the product $s_{\bar p}\big[\pi^*f\star(\pi^*g\star\pi^*h)\big]$ we observe that there is an obvious one-to-one correspondence between differential operators ∂_i on $\mathcal{A}=C^\infty(\mathcal{M})$ and on $A=C^\infty(M)$, and that for any function f on M one has $s_{\bar p}(\partial_i(\pi^*f))=\partial_if$. This implies that the associator Φ naturally acts as a differential operator on $A\otimes A\otimes A$. We define

$$\Phi^{-1}(\,\underline{f}\otimes\underline{g}\otimes\underline{h}\,)=\underline{f}^{\bar\phi}\otimes\underline{g}^{\bar\phi}\otimes\underline{h}^{\bar\phi}\qquad\text{and}\qquad\Phi^{-1}(f\otimes g\otimes h)=f^{\bar\phi}\otimes g^{\bar\phi}\otimes h^{\bar\phi} \tag{3.34}$$

with implicit summation as before, and then we find

$$\begin{aligned}
s_{\bar p}^*\big[\big(\pi^*f\star(\pi^*g\star\pi^*h)\big)\big] &= s_{\bar p}^*\big[\big((\pi^*f)^{\bar\phi}\star(\pi^*g)^{\bar\phi}\big)\star(\pi^*h)^{\bar\phi}\,\big]\\
&= s_{\bar p}^*\big[\big((\pi^*f^{\bar\phi})\star(\pi^*g^{\bar\phi})\big)\star(\pi^*h^{\bar\phi}\,)\big]\\
&= \mu_{\bar p}^{(3)}\big(f^{\bar\phi}\otimes g^{\bar\phi}\otimes h^{\bar\phi}\,\big)\\
&= \star_{\bar p}\Big(\exp\big(\tfrac{\ell_s^4}{12}\,R^{ijk}\,\partial_i\otimes\partial_j\otimes\partial_k\big)\,\Phi^{-1}(f\otimes g\otimes h)\Big)\\
&= \star_{\bar p}\Big(\exp\big(-\tfrac{\ell_s^4}{12}\,R^{ijk}\,\partial_i\otimes\partial_j\otimes\partial_k\big)(f\otimes g\otimes h)\Big)\ .
\end{aligned}\tag{3.35}$$

Thus we can generate the new triproduct $s_{\bar p}^*\big[\big(\pi^*f\star(\pi^*g\star\pi^*h)\big)\big]$ of fields by applying the inverse of the associator Φ directly to functions on M as in (3.35). This new triproduct also obeys Proposition 3.19, since from (3.35) we have

$$\int_M\mathrm{d}^dx\ s_{\bar p}^*\big[\big(\pi^*f\star(\pi^*g\star\pi^*h)\big)\big]=\int_M\mathrm{d}^dx\ \mu_{\bar p}^{(3)}\big[\Phi^{-1}(f\otimes g\otimes h)\big]=\int_M\mathrm{d}^dx\ f\star_{\bar p}g\star_{\bar p}h\ . \tag{3.36}$$

A quick way to prove this is to note that

$$\Phi^{-1}\big(\mathrm{e}^{\,\mathrm{i}k_1\cdot x}\otimes\mathrm{e}^{\,\mathrm{i}k_2\cdot x}\otimes\mathrm{e}^{\,\mathrm{i}k_3\cdot x}\big)=\mathrm{e}^{\,\frac{\mathrm{i}\,\ell_s^4}{6}\,R(k_1,k_2,k_3)}\,\big(\mathrm{e}^{\,\mathrm{i}k_1\cdot x}\otimes\mathrm{e}^{\,\mathrm{i}k_2\cdot x}\otimes\mathrm{e}^{\,\mathrm{i}k_3\cdot x}\big)\ , \tag{3.37}$$

and the extra phase factor is unity after imposing momentum conservation $k_1+k_2+k_3=0$.

We can now generate all induced 4-triproducts from the diagram (3.31). For example one has

$$\begin{aligned}
s_{\bar p}^*\big[(\pi^*f\star\pi^*g)\star(\pi^*h\star\pi^*k)\big] &= \mu_{\bar p}^{(4)}\big[\Phi^{-1}_{A\otimes A,A,A}\big((f\otimes g)\otimes h\otimes k\big)\big]\\
&= \mu_{\bar p}^{(4)}\big[\exp\big(-\tfrac{\ell_s^4}{6}\,R^{ijk}\,(\partial_i^1\,\partial_j^3\,\partial_k^4+\partial_i^2\,\partial_j^3\,\partial_k^4)\big)(f\otimes g\otimes h\otimes k)\big]\\
&= \star_{\bar p}\big[\exp\big(\tfrac{\ell_s^4}{12}\,R^{ijk}\,(\partial_i^1\,\partial_j^2\,\partial_k^3+\partial_i^1\,\partial_j^2\,\partial_k^4\\
&\qquad -\partial_i^1\,\partial_j^3\,\partial_k^4-\partial_i^2\,\partial_j^3\,\partial_k^4)\big)(f\otimes g\otimes h\otimes k)\big]\ .
\end{aligned}\tag{3.38}$$

A completely analogous calculation following both upper arrows in the diagram (3.31) gives

$$s_{\bar p}^*\big[\pi^*f\star\big(\pi^*g\star(\pi^*h\star\pi^*k)\big)\big]\ =\ \mu_{\bar p}^{(4)}\big[\Phi^{-1}_{A\otimes A,A,A}\,\Phi^{-1}_{A,A,A\otimes A}\big(f\otimes g\otimes(h\otimes k)\big)\big] \tag{3.39}$$

$$= \star_{\bar p}\Big[\exp\Big(-\frac{\ell_s^4}{12}\,R^{ijk}\sum_{1\le a<b<c\le 4}\partial_i^a\,\partial_j^b\,\partial_k^c\Big)(f\otimes g\otimes h\otimes k)\Big]\ .$$

These new 4-triproducts serve just as well for describing the products among off-shell closed string tachyon vertex operators, and indeed we find as in Proposition 3.19 the on-shell result

$$\begin{aligned}\int_M \mathrm{d}^dx\ s_{\bar p}^*\big[(\pi^*f\star\pi^*g)\star(\pi^*h\star\pi^*k)\big] &= \int_M \mathrm{d}^dx\ \mu_{\bar p}^{(4)}\big[\Phi^{-1}_{A\otimes A,\,A,\,A}\big((f\otimes g)\otimes h\otimes k\big)\big]\\ &= \int_M \mathrm{d}^dx\ \star_{\bar p}\big[\Phi^{-1}_{A\otimes A,\,A,\,A}\big((f\otimes g)\otimes h\otimes k\big)\big]\\ &= \int_M \mathrm{d}^dx\ \star_{\bar p}\big[\Phi^{-1}\big((f\star_{\bar p} g)\otimes h\otimes k\big)\big]\\ &= \int_M \mathrm{d}^dx\ f\star_{\bar p} g\star_{\bar p} h\star_{\bar p} k \qquad (3.40)\end{aligned}$$

where in the last equality we used (3.36). Letting $R^{ijk}\to -R^{ijk}$ in Proposition 3.19 we also immediately see that the integral of (3.39) coincides with (3.40).

The situation is notably different if one follows the left vertical arrow in the diagram (3.31). We first compute the product

$$\begin{aligned}s_{\bar p}^*\big[\big(\pi^*f\star(\pi^*g\star\pi^*h)\big)\star\pi^*k\big] &= \mu_{\bar p}^{(4)}\big[\Phi^{-1}_{A,\,A\otimes A,\,A}\big(f\otimes(g\otimes h)\otimes k\big)\big] \qquad (3.41)\\ &= \star_{\bar p}\Big[\exp\big(\tfrac{\ell_s^4}{12}\,R^{ijk}\,(-\partial_i^1\,\partial_j^2\,\partial_k^3+\partial_i^1\,\partial_j^2\,\partial_k^4\\ &\qquad +\partial_i^1\,\partial_j^3\,\partial_k^4+\partial_i^2\,\partial_j^3\,\partial_k^4)\big)(f\otimes g\otimes h\otimes k)\Big]\ .\end{aligned}$$

Analogously we find for the product in the bottom right corner of (3.31) as

$$\begin{aligned}s_{\bar p}^*\big[\pi^*f\star\big((\pi^*g\star\pi^*h)\star\pi^*k\big)\big] &= \mu_{\bar p}^{(4)}\big[(\Phi^{-1}\otimes\mathrm{id})\,\Phi^{-1}_{A,\,A\otimes A,\,A}\big(f\otimes(g\otimes h)\otimes k\big)\big] \qquad (3.42)\\ &= \star_{\bar p}\Big[\exp\big(\tfrac{\ell_s^4}{12}\,R^{ijk}\,(-\partial_i^1\,\partial_j^2\,\partial_k^3-\partial_i^1\,\partial_j^2\,\partial_k^4\\ &\qquad -\partial_i^1\,\partial_j^3\,\partial_k^4+\partial_i^2\,\partial_j^3\,\partial_k^4)\big)(f\otimes g\otimes h\otimes k)\Big]\ ,\end{aligned}$$

and it is easy to see that a further application of $\mathrm{id}\otimes\Phi^{-1}$ leads exactly to the 4-triproduct (3.39). As previously the application of $\Phi^{-1}_{A,\,A\otimes A,\,A}$ does not alter closed string amplitudes, but the application of $\Phi^{-1}\otimes\mathrm{id}$ does, and we find

$$\begin{aligned}\int_M \mathrm{d}^dx\ s_{\bar p}^*\big[\big(\pi^*f\star(\pi^*g\star\pi^*h)\big)\star\pi^*k\big] &= \int_M \mathrm{d}^dx\ \mu_{\bar p}^{(4)}\big[\Phi^{-1}(f\otimes g\otimes h)\otimes k\big] \qquad (3.43)\\ &= \int_M \mathrm{d}^dx\ \mu_{\bar p}^{(4)}\big(f^\phi\otimes g^\phi\otimes h^\phi\otimes k\big)\\ &= \int_M \mathrm{d}^dx\ s_{\bar p}^*\big[\pi^*f\star\big((\pi^*g\star\pi^*h)\star\pi^*k\big)\big]\ .\end{aligned}$$

In general one has

$$\int_M \mathrm{d}^dx\ \mu_{\bar p}^{(4)}\big(f^{\bar\phi}\otimes g^{\bar\phi}\otimes h^{\bar\phi}\otimes k\big) = \int_M \mathrm{d}^dx\ f^{\bar\phi}\star_{\bar p} g^{\bar\phi}\star_{\bar p} h^{\bar\phi}\star_{\bar p} k$$

$$\neq \int_M \mathrm{d}^d x \; f \star_{\bar p} g \star_{\bar p} h \star_{\bar p} k \ . \tag{3.44}$$

As a consequence these two 4-triproducts leave traces of nonassociativity through additional interaction terms, and should hence not be considered as physically viable products among off-shell closed string tachyon vertex operators.

This line of reasoning can be extended to all n-triproducts with $n > 4$. By MacLane's coherence theorem, all possible bracketings of star products of n functions on $\mathcal{M}$ are related by successive applications of the inverse associator Φ^{-1} to tensor products of these n functions. In all there are C_{n-1} star products, where C_n is the Catalan number of degree n. Of these there are $n-1$ triproducts which can serve as physical products among closed string vertex operators. They are obtained from the bracketings

$$\big((\cdots((\underline{f}_1 \star \underline{f}_2) \star \underline{f}_3) \star \cdots) \star \underline{f}_r\big) \star \big(\underline{f}_{r+1} \star (\underline{f}_{r+2} \star (\underline{f}_{r+3} \star (\cdots \star \underline{f}_n) \cdots))\big) \tag{3.45}$$

for $1 \leq r \leq n-1$, which for $r \leq n-2$ are obtained by successive applications for $s = r, r+1, \dots, n-3$ of the inverse associators

$$\Phi^{-1}_{(\cdots((\mathcal{A}\otimes\mathcal{A})\otimes\mathcal{A})\otimes\cdots)\otimes\mathcal{A}\,,\,\mathcal{A}\,,\,\mathcal{A}\otimes(\mathcal{A}\otimes(\mathcal{A}\otimes(\cdots\otimes\mathcal{A})\cdots))} \tag{3.46}$$

to $((\cdots(\underline{f}_1 \otimes \underline{f}_2) \otimes \underline{f}_3) \otimes \cdots) \otimes \underline{f}_r) \otimes \underline{f}_{r+1} \otimes (\underline{f}_{r+2} \otimes (\underline{f}_{r+3} \otimes (\cdots \otimes \underline{f}_n) \cdots)))$, where the first slot of Φ^{-1} in (3.46) contains s tensor products of the algebra $\mathcal{A} = C^\infty(\mathcal{M})$. The action of the inverse associator (3.46) is again obtained using the Leibniz rule for the partial derivatives ∂_i on $\mathcal{A}^{\otimes s}$ and $\mathcal{A}^{\otimes n-s-1}$. It follows that for the specific associator we are considering there is no ambiguity in rewriting the expression (3.46) simply as $\Phi^{-1}_{\mathcal{A}^{\otimes s},\,\mathcal{A}\,,\,\mathcal{A}^{\otimes n-s-1}}$. Its momentum space representation is obtained by applying it to the tensor products of plane waves

$$\begin{aligned}\big[(\cdots(\mathrm{e}^{\,\mathrm{i}\,k_1\cdot x} \otimes \mathrm{e}^{\,\mathrm{i}\,k_2\cdot x}) \otimes \cdots) \otimes \mathrm{e}^{\,\mathrm{i}\,k_s\cdot x}\big] \ \otimes \ \mathrm{e}^{\,\mathrm{i}\,k_{s+1}\cdot x} \\ \otimes \ \big[\mathrm{e}^{\,\mathrm{i}\,k_{s+2}\cdot x} \otimes (\mathrm{e}^{\,\mathrm{i}\,k_{b+3}\cdot x} \otimes (\cdots \otimes \mathrm{e}^{\,\mathrm{i}\,k_n\cdot x}) \cdots)\big]\end{aligned} \tag{3.47}$$

which yields the phase factor

$$\exp\big(\tfrac{\mathrm{i}\,\ell_s^4}{6}\, R(k_1 + \cdots + k_s\,,\, k_{s+1}\,,\, k_{s+2} + \cdots + k_n)\big) \ . \tag{3.48}$$

The n-triproducts (3.45) on phase space $\mathcal{M}$ induce triproducts on configuration space M by setting $\underline{f}_i = \pi^* f_i$ for $i = 1, \dots, n$ and then pulling these products back along the local sections $s_{\bar p} : M \to \mathcal{M}$. Their explicit expressions in terms of the Moyal-Weyl product and of the R-flux tridifferential operators is obtained by replacing the tridifferential operators of Proposition 3.9 with

$$\mathrm{e}^{\,R_{(r,n)}} := \exp\Big(\frac{\ell_s^4}{12} \sum_{1\leq a<b<c\leq n} R^{ijk}\,\partial_i^a\,\partial_j^b\,\partial_k^c \ - \ \frac{\ell_s^4}{6} \sum_{b=r+1}^{n-1} \ \sum_{1\leq a<b<c\leq n} R^{ijk}\,\partial_i^a\,\partial_j^b\,\partial_k^c\Big) \ . \tag{3.49}$$

These n-triproducts also obey the on-shell condition of Proposition 3.19 (as can be seen from the vanishing of the phase in (3.48) after integrating over M which yields momentum conservation $k_1 + \cdots + k_n = 0$).

The remaining triproducts violate the on-shell condition because they are obtained from the previous ones by applying those inverse associators that like the vertical arrows in (3.31) do not act on all the tensor products entries; in this case total momentum conservation does not imply the trivialization of their action. For example one has

$$\int_M \mathrm{d}^d x\ s_{\bar p}^*\Big[\big(\big((\cdots((\pi^*f_1\star\pi^*f_2)\star\pi^*f_3)\star\cdots)\star\pi^*f_{r-2}\big)\star(\pi^*f_{r-1}\star\pi^*f_r)\big)$$
$$\star\big(\pi^*f_{r+1}\star(\pi^*f_{r+2}\star(\pi^*f_{r+3}\star(\cdots\star\pi^*f_n)\cdots))\big)\Big]$$
$$=\int_M \mathrm{d}^d x\ \star_{\bar p}\Big[\mathrm{e}^{\,R_{(r,n)}}\,\big(\Phi^{-1}_{\mathcal{A}^{\otimes r-2},\mathcal{A},\mathcal{A}}\otimes\mathrm{id}^{\otimes n-r}\big)(f_1\otimes\cdots\otimes f_n)\Big]$$
$$=\int_M \mathrm{d}^d x\ \star_{\bar p}\Big[\Phi^{-1}_{\mathcal{A}^{\otimes r-2},\mathcal{A},\mathcal{A}}(f_1\otimes\cdots\otimes f_r)\Big]\star_{\bar p} f_{r+1}\star_{\bar p} f_{r+2}\cdots\star_{\bar p} f_n\ . \tag{3.50}$$

Because of momentum conservation, all n-triproducts which differ from that above by a sequence of inverse associators acting non-trivially on each tensor product entry (like the horizontal or diagonal arrows in (3.31)) yield the same integral (3.50). This procedure can be iterated to obtain all remaining triproducts and their on-shell associativity violating amplitudes, but we refrain from detailing further the combinatorics.

4. Triproducts from phase space cochains

4.1. Motivation: Differential geometry of nonassociative R-space

In order to extend the considerations of Section 3 to more general geometric entities, such as differential forms and vector fields, we need to uncover the more algebraic construction that underlies the explicit representation via Fourier transformation that we used so far. Such a description of the phase space nonassociative geometry was introduced by [27], while in [7] a general categorical perspective on cochain twist deformation is presented. Here we shall further develop the phase space nonassociative geometry, recover the main results of Section 3 in this algebraic context, and then study the exterior product of forms in nonassociative phase space; we further derive the induced products on the leaves of constant momentum and their property under integration. We also study the deformed Lie algebra of infinitesimal diffeomorphisms on phase space, as well as its action on exterior forms and tensor fields. These geometric structures are again induced on each constant momentum leaf and in particular in configuration space (the leaf of vanishing momentum). We thus derive the configuration space geometry of exterior forms and vector fields, together with their deformed Lie algebra and Jacobiator, at all orders in the R-flux components R^{ijk} from the geometry of phase space that we canonically construct via 2-cochain twist deformation.

4.2. Cochain twist deformations and nonassociative star products

Consider the Lie algebra $\mathfrak{h}$ with generators $P_i, \tilde P^i, M^i$ for $i=1,\dots,d$ and relations

$$[\tilde P^i, M^j] = \tfrac{1}{6}\,\ell_s^4\,R^{ijk}\,P_k\ , \tag{4.1}$$

with all other Lie brackets vanishing. This Lie algebra naturally acts on functions on phase space via the representation

$$P_i(\,\underline{f}\,) := \partial_i\,\underline{f}\ ,\qquad \tilde P^i(\,\underline{f}\,) := \mathrm{i}\,\hbar\,\tilde\partial^i\,\underline{f}\qquad\text{and}\qquad M^i(\,\underline{f}\,) = \frac{\mathrm{i}\,\ell_s^4}{6\hbar}\,R^{ijk}\,p_j\,\partial_k\,\underline{f} \tag{4.2}$$

for all $\underline{f} \in C^\infty(\mathcal{M})$. The operators P_i and $\tilde{P}^i$ respectively generate position and momentum translations in phase space, while M^i generate Bopp shifts. Alternatively, we can represent $\mathfrak{h}$ on the algebra $\mathsf{Diff}(M)$ of differential operators on M by

$$P_i = ad_{\partial_i} \ , \qquad \tilde{P}^i = ad_{x^i} \qquad \text{and} \qquad M^i = \tfrac{1}{6}\,\ell_s^4\, R^{ijk}\,\partial_j\, ad_{\partial_k} \ , \tag{4.3}$$

with $ad_{\partial_i}(x^j) = \delta_i^j = -ad_{x^j}(\partial_i)$.

Let $U(\mathfrak{h})$ be the universal enveloping algebra of the Lie algebra $\mathfrak{h}$, i.e., the associative algebra of $\mathbb{C}$-linear combinations of products of the generators $P_i, \tilde{P}^i, M^i$ modulo the Lie algebra relations (4.1) as well as all other vanishing relations $[P_i, P_j] = 0, [P_i, M^j] = 0, [M^i, M^j] = 0$, etc. This associative unital algebra is a Hopf algebra with coproduct $\Delta : U(\mathfrak{h}) \to U(\mathfrak{h}) \otimes U(\mathfrak{h})$, counit $\varepsilon : U(\mathfrak{h}) \to \mathbb{C}$ and antipode $S : U(\mathfrak{h}) \to U(\mathfrak{h})$ defined on generators $X \in \{P_i, \tilde{P}^i, M^i\}$ as

$$\Delta(X) = X \otimes 1 + 1 \otimes X \ , \qquad \varepsilon(X) = 0 \qquad \text{and} \qquad S(X) = -X \ . \tag{4.4}$$

The coproduct Δ and counit ε are linear maps extended multiplicatively to all of $U(\mathfrak{h})$, while the antipode S is a linear map extended anti-multiplicatively to all of $U(\mathfrak{h})$, i.e., $S(X\,Y) = S(Y)\,S(X)$. The action of $\mathfrak{h}$ on $C^\infty(\mathcal{M})$ naturally extends to an action of $U(\mathfrak{h})$ on $C^\infty(\mathcal{M})$ via

$$(X_1\, X_2 \cdots X_n)(\,\underline{f}\,) := X_1\big(X_2(\cdots X_n(\,\underline{f}\,)\cdots)\big) \tag{4.5}$$

for all $X_1, X_2, \dots, X_n \in \mathfrak{h}$ and $\underline{f} \in C^\infty(\mathcal{M})$.

The coproduct on generators X encodes the Leibniz rule $X(\,\underline{f}\ \underline{g}\,) = X(\,\underline{f}\,)\underline{g} + \underline{f}\,X(\,\underline{g}\,)$ for all functions $\underline{f}\,,\,\underline{g}$ on phase space. An equivalent expression for the Leibniz rule is $X(\,\underline{f}\ \underline{g}\,) = \mu \circ \Delta(X)(\,\underline{f} \otimes \underline{g}\,)$. Multiplicativity of the coproduct implies more generally that

$$\zeta(\,\underline{f}\ \underline{g}\,) = \mu\big(\Delta(\zeta)(\,\underline{f} \otimes \underline{g}\,)\big) \tag{4.6}$$

for all $\zeta \in U(\mathfrak{h})$, which equivalently reads $\zeta \circ \mu\,(\,\underline{f} \otimes \underline{g}\,) = \mu \circ \Delta(\zeta)(\,\underline{f} \otimes \underline{g}\,)$.

We can consider two twists in $U(\mathfrak{h}) \otimes U(\mathfrak{h})$ given by[2]

$$F = \exp\big(-\tfrac{1}{2}\,(P_i \otimes \tilde{P}^i - \tilde{P}^i \otimes P_i)\big) \qquad \text{and} \qquad F' = \exp\big(-\tfrac{1}{2}\,(M^i \otimes P_i - P_i \otimes M^i)\big) \ . \tag{4.7}$$

They are both abelian cocycle twists of the Hopf algebra $U(\mathfrak{h})$ with the coalgebra structures Δ, ε, S, i.e., they are invertible and satisfy the relations

$$(F \otimes 1)\,(\Delta \otimes \mathrm{id})F \ = \ (1 \otimes F)\,(\mathrm{id} \otimes \Delta)F \ , \tag{4.8}$$

$$(\varepsilon \otimes \mathrm{id})F \ = \ 1 \ = \ (\mathrm{id} \otimes \varepsilon)F \ , \tag{4.9}$$

plus the analogous relations for $F \to F'$. The first relation is a 2-cocycle condition which assures that the star products obtained from the twists are associative. The second relation is just a normalization condition. Since the bracket (4.1) is central in the Lie algebra $\mathfrak{h}$, we can use the Baker–Campbell–Hausdorff formula

$$\exp(A)\ \exp(B) = \exp\big([A, B]\big)\ \exp(B)\ \exp(A) \tag{4.10}$$

[2] The twists F and F' should be more precisely regarded respectively as power series expansions in $\hbar$ and $\ell_s^4/\hbar$, the deformation parameters that for ease of notation have been absorbed into the definition of the generators $\tilde{P}^i$ and M^i respectively. Then $U(\mathfrak{h})$ is an algebra over $\mathbb{C}[[\hbar, \ell_s^4/\hbar]]$, the formal power series in $\hbar$ and $\ell_s^4/\hbar$ with coefficients in $\mathbb{C}$. In this setting the twists F and F' live in a completion of the tensor product $U(\mathfrak{h}) \otimes U(\mathfrak{h})$.

for $[A,B]$ central (and A,B in $U(\mathfrak{h})\otimes U(\mathfrak{h})$ and then in $U(\mathfrak{h})\otimes U(\mathfrak{h})\otimes U(\mathfrak{h})$) to compute

$$F\,F' = F'\,F \qquad \text{and} \qquad (1\otimes F')\,(\mathrm{id}\otimes\Delta)F = \mathrm{e}^{\,R}\,(\mathrm{id}\otimes\Delta)F\,(1\otimes F') \tag{4.11}$$

where we have introduced the central element in $U(\mathfrak{h})\otimes U(\mathfrak{h})\otimes U(\mathfrak{h})$ given by

$$\mathrm{e}^{\,R} := \exp\big(\tfrac{\ell_s^4}{12}\,R^{ijk}\,P_i\otimes P_j\otimes P_k\big)\ . \tag{4.12}$$

Note that, in the representation (4.2) on $C^\infty(\mathcal{M})$, the square of this operator is the associator (3.25), i.e.,

$$\Phi = \mathrm{e}^{\,2R}\ . \tag{4.13}$$

The n-triproducts of functions from Section 3 can also be obtained from more algebraic considerations using the twists F and F' of the Hopf algebra $U(\mathfrak{h})$. On one hand this requires a minimal amount of Hopf algebra technology, while on the other hand this approach can be applied to any algebra that carries a representation of $U(\mathfrak{h})$. In particular, we shall apply it to the algebras of exterior differential forms and of Lie derivatives (infinitesimal diffeomorphisms).

Let us first consider the twist F'. Associated with F' there is a new Hopf Algebra $U(\mathfrak{h})^{F'}$. The algebra structure is the same as that of $U(\mathfrak{h})$, the new coproduct is given by

$$\Delta^{F'}(\xi) = F'\,\Delta(\xi)\,F'^{-1} \tag{4.14}$$

for all $\xi\in U(\mathfrak{h})$, while $U(\mathfrak{h})^{F}$ has the same counit ε and antipode S as $U(\mathfrak{h})$ due to the abelian structure of the twist, i.e., P_i and M^j commute. We can now further deform the Hopf algebra $U(\mathfrak{h})^{F'}$ with the twist F. Notice that while F is a cocycle twist for the Hopf algebra $U(\mathfrak{h})$, because it satisfies (4.8) and (4.9), it is *not* a cocycle twist of the new Hopf algebra $U(\mathfrak{h})^{F'}$. To compute its failure it is convenient to compare the actions of the coproducts $\Delta^{F'}$ and Δ on the twist element F.

Proposition 4.15. $(\mathrm{id}\otimes\Delta^{F'})F = \mathrm{e}^{\,R}\,(\mathrm{id}\otimes\Delta)F$ *and* $(\Delta^{F'}\otimes\mathrm{id})F = \mathrm{e}^{\,-R}\,(\Delta\otimes\mathrm{id})F$.

Proof. The coproduct $\Delta^{F'}$ differs from Δ when applied to the generators $\tilde P^i$. We consider the term in (4.10) which is linear in B, i.e., $\exp(A)\,B\,\exp(-A) = B+[A,B]$, and we identify B with $\Delta(\tilde P^i)$ in order to easily compute

$$\begin{aligned}\Delta^{F'}(\tilde P^i) &= F'\,\Delta(\tilde P^i)\,F'^{-1}\\ &= F'\,\big(\tilde P^i\otimes 1+1\otimes\tilde P^i\big)\,F'^{-1}\\ &= \tilde P^i\otimes 1+1\otimes\tilde P^i-\tfrac12\,\big[M^k\otimes P_k-P_k\otimes M^k\,,\,\tilde P^i\otimes 1+1\otimes\tilde P^i\big]\\ &= \Delta(\tilde P^i)-\tfrac16\,\ell_s^4\,R^{ijk}\,P_j\otimes P_k\ .\end{aligned} \tag{4.16}$$

Next we compute

$$\begin{aligned}(\mathrm{id}\otimes\Delta^{F'})F &= (\mathrm{id}\otimes\Delta^{F'})\exp\big(-\tfrac12\,(P_i\otimes\tilde P^i-\tilde P^i\otimes P_i)\big)\\ &= \exp\Big(-\tfrac12\,\big(P_i\otimes\Delta^{F'}(\tilde P^i)-\tilde P^i\otimes\Delta^{F'}(P_i)\big)\Big)\\ &= \exp\Big(-\tfrac12\,\big(P_i\otimes\Delta(\tilde P^i)-\tilde P^i\otimes\Delta(P_i)\big)\Big)\,\exp\big(\tfrac{\ell_s^4}{12}\,R^{ijk}\,P_i\otimes P_j\otimes P_k\big)\\ &= (\mathrm{id}\otimes\Delta)F\,\mathrm{e}^{\,R} \;=\; \mathrm{e}^{\,R}\,(\mathrm{id}\otimes\Delta)F\ ,\end{aligned} \tag{4.17}$$

where in the second equality we used multiplicativity of the coproduct Δ. The proof of the second identity is very similar, or alternatively observe that $F \to F^{-1}$ under flipping of the order of its legs in the tensor product $U(\mathfrak{h}) \otimes U(\mathfrak{h})$. □

As a corollary we find that F is not a cocycle twist of $U(\mathfrak{h})^{F'}$, i.e., it fails the cocycle condition for the Hopf algebra $U(\mathfrak{h})^{F'}$.

Corollary 4.18. $(1\otimes F)\big(\mathrm{id}\otimes\Delta^{F'}\big)F = \mathrm{e}^{2R}\,(F\otimes 1)\,\big(\Delta^{F'}\otimes\mathrm{id}\big)F$.

Proof. Use Proposition 4.15 to rewrite the left-hand and right-hand sides in terms of the undeformed coproduct Δ. Then recall that e^{R} is central in the algebra $U(\mathfrak{h})\otimes U(\mathfrak{h})\otimes U(\mathfrak{h})$, and finally use the cocycle property (4.8). □

The Hopf algebra $U(\mathfrak{h})$ acts on the algebra $C^\infty(\mathcal{M})$ of functions on phase space via the representation (4.2), and this action is compatible with the product in $C^\infty(\mathcal{M})$ in the sense of (4.6). Because of this compatibility we can then deform $C^\infty(\mathcal{M})$ into a new algebra $C^\infty(\mathcal{M})_{\star_{F'}}$ defined by the new product

$$\underline{f} \star_{F'} \underline{g} = \mu_{F'}(\underline{f}\otimes\underline{g}) := \mu\big(F'^{-1}(\underline{f}\otimes\underline{g})\big) \tag{4.19}$$

for any phase space functions $\underline{f}$ and $\underline{g}$. The algebra $C^\infty(\mathcal{M})_{\star_{F'}}$ is noncommutative but associative because of the cocycle condition satisfied by F'. On $C^\infty(\mathcal{M})_{\star_{F'}}$ there is a natural action of the Hopf algebra $U(\mathfrak{h})^{F'}$; it is again defined by (4.2). The deformed coproduct $\Delta^{F'}$, which characterizes $U(\mathfrak{h})^{F'}$, describes the action of any element $\zeta\in U(\mathfrak{h})^{F'}$ on products of functions, i.e., for all $\underline{f}, \underline{g}\in C^\infty(\mathcal{M})_{\star_{F'}}$, by (4.6) one has

$$\begin{aligned}\zeta(\underline{f}\star_{F'}\underline{g}) &= \zeta\circ\mu_{F'}(\underline{f}\otimes\underline{g})\\ &= \zeta\circ\mu\circ F'^{-1}(\underline{f}\otimes\underline{g})\\ &= \mu\circ\Delta(\zeta)\circ F'^{-1}(\underline{f}\otimes\underline{g}) \;=\; \mu_{F'}\circ\Delta^{F'}(\zeta)(\underline{f}\otimes\underline{g})\,.\end{aligned} \tag{4.20}$$

In particular it leads to a deformed Leibniz rule for the generators $\tilde{P}^i$ as can be read off from (4.16).

Now we iterate this procedure and deform the Hopf algebra $U(\mathfrak{h})^{F'}$ and the noncommutative algebra $C^\infty(\mathcal{M})_{\star_{F'}}$ with the 2-cochain $F\in U(\mathfrak{h})^{F'}\otimes U(\mathfrak{h})^{F'}$. This is equivalent to the deformation of the original Hopf algebra $U(\mathfrak{h})$ and the algebra of smooth functions $C^\infty(\mathcal{M})$ with the 2-cochain

$$\mathcal{F} := F\,F' = F'\,F\,. \tag{4.21}$$

Since F fails the $\Delta^{F'}$ 2-cocycle property, the coproduct $(\Delta^{F'})^F$, defined as

$$(\Delta^{F'})^F(\zeta) = F\,\Delta^{F'}(\zeta)\,F^{-1} = \Delta^{\mathcal{F}}(\zeta)\,, \tag{4.22}$$

will equip $(U(\mathfrak{h})^{F'})^F = U(\mathfrak{h})^{\mathcal{F}}$ with the structure of a quasi-Hopf algebra [16] (see also [27, 7]). Correspondingly, $C^\infty(\mathcal{M})_{\star_{F'}}$ deforms to a quasi-associative noncommutative algebra with product

$$\underline{f}\star_{\mathcal{F}}\underline{g} = \underline{f}\star_{F\,F'}\underline{g} = \mu_{F'}\big(F^{-1}(\underline{f}\otimes\underline{g})\big) = \mu\big((F\,F')^{-1}(\underline{f}\otimes\underline{g})\big) \tag{4.23}$$

for all phase space functions $\underline{f}, \underline{g}$. The failure of associativity of $\star_{\mathcal{F}}$ is due to the failure of the cocycle condition for F: We have

$$(\underline{f}\star_{F\,F'}\underline{g})\star_{F\,F'}\underline{h} \;=\; \mu_{F\,F'}\circ(\mu_{F\,F'}\otimes\mathrm{id})(\underline{f}\otimes\underline{g}\otimes\underline{h})$$

$$
\begin{aligned}
&= \mu_{F'} \circ F^{-1} \circ (\mu_{F'} \otimes 1) \circ (F^{-1} \otimes \mathrm{id})(\underline{f} \otimes \underline{g} \otimes \underline{h}) \\
&= \mu_{F'} \circ (\mu_{F'} \otimes \mathrm{id}) \circ (\Delta^{F'} \otimes \mathrm{id}) F^{-1} (F^{-1} \otimes 1)(\underline{f} \otimes \underline{g} \otimes \underline{h}) \qquad (4.24) \\
&= \mu_{F'} \circ (\mathrm{id} \otimes \mu_{F'}) \circ (\mathrm{id} \otimes \Delta^{F'}) F^{-1} (1 \otimes F^{-1})\, \mathrm{e}^{2R} (\underline{f} \otimes \underline{g} \otimes \underline{h})
\end{aligned}
$$

where in the third line we used (4.20) on the first tensor product entry of F^{-1}, while in the fourth line we used associativity of the $\star_{F'}$ product in the form $\mu_{F'} \circ (\mu_{F'} \otimes \mathrm{id}) = \mu_{F'} \circ (\mathrm{id} \otimes \mu_{F'})$ together with the inverse of the 2-cochain property of Corollary 4.18. Similarly, we have

$$
\begin{aligned}
\underline{f} \star_{F\,F'} (\underline{g} \star_{F\,F'} \underline{h}) &= \mu_{F\,F'} \circ (\mathrm{id} \otimes \mu_{F\,F'})(\underline{f} \otimes \underline{g} \otimes \underline{h}) \\
&= \mu_{F'} \circ (\mathrm{id} \otimes \mu_{F'})(\mathrm{id} \otimes \Delta^{F'}) F^{-1} (1 \otimes F^{-1})(\underline{f} \otimes \underline{g} \otimes \underline{h}) \ . \qquad (4.25)
\end{aligned}
$$

Comparison of these products gives (3.28).

For later use, we note that we have chosen to derive the nonassociativity structure of the $\star_{\mathcal{F}}$ product using associativity of the $\star_{F'}$ product and the 2-cochain property of Corollary 4.18. We could equally as well have decomposed the $\star_{\mathcal{F}}$ product as $\mu_{\mathcal{F}} = \mu \circ \mathcal{F}^{-1}$ and then, following similar steps as in (4.24), used associativity of the undeformed product μ and the 2-cochain property for $\mathcal{F} = F\,F'$ (or for $\mathcal{F}^{-1}$) which is not difficult to show to be given by

$$
(\Delta \otimes \mathrm{id}) \mathcal{F}^{-1} (\mathcal{F}^{-1} \otimes 1) = (\mathrm{id} \otimes \Delta) \mathcal{F}^{-1} (1 \otimes \mathcal{F}^{-1})\, \mathrm{e}^{2R} \ . \qquad (4.26)
$$

In a similar vein, using the isomorphism $C^\infty(\mathcal{M})_{\star_F} \cong \mathsf{Diff}(M)$ of noncommutative associative algebras and the representation (4.3) of the Hopf algebra $U(\mathfrak{h})$, one can deform the algebra of differential operators on configuration space M to a quasi-associative noncommutative algebra $\mathsf{Diff}(M)_{\star_{F'}}$.

4.3. Configuration space triproducts

Let us now demonstrate how to reproduce the triproducts of Section 3 within the present formalism. The pullback of the product on the leaf of constant momentum $\bar{p}$ is given by

$$
\begin{aligned}
s_{\bar{p}}^*(\underline{f} \star_{F\,F'} \underline{g}) &= \mu_{F'}\big(F^{-1}(s_{\bar{p}}^*\underline{f} \otimes s_{\bar{p}}^*\underline{g})\big) \\
&= \mu_{F'}(s_{\bar{p}}^*\underline{f} \otimes s_{\bar{p}}^*\underline{g}) = \mu\big(\exp\big(\tfrac{\mathrm{i}\,\ell_s^4}{6\hbar}\,\theta_{\bar{p}}\big)(s_{\bar{p}}^*\underline{f} \otimes s_{\bar{p}}^*\underline{g})\big) \qquad (4.27)
\end{aligned}
$$

where in the second equality we observed that F acts as the identity on functions of constant momentum, while in the third equality we recalled the definition of the bivector $\theta_{\bar{p}}$ from (3.4). In particular if f, g are functions on configuration space and $\underline{f} = \pi^* f$, $\underline{g} = \pi^* g$, then we recover the 2-product $\mu_{\bar{p}}^{(2)}$ from (3.6),

$$
s_{\bar{p}}^*(\pi^* f \star_{F\,F'} \pi^* g) = f \star_{\bar{p}} g = \mu_{\bar{p}}^{(2)}(f \otimes g) \ . \qquad (4.28)
$$

We now proceed to the product of three phase space functions and its pullback on the leaf of constant momentum $\bar{p}$. We substitute in the triple product expression (4.24) the inverse of the second identity of Proposition 4.15 in order to express $\Delta^{F'}$ in terms of Δ and obtain

$$
(\underline{f} \star_{F\,F'} \underline{g}) \star_{F\,F'} \underline{h} = \mu_{F'} \circ (\mu_{F'} \otimes \mathrm{id}) \circ \mathrm{e}^{R} (\Delta \otimes \mathrm{id}) F^{-1} (F^{-1} \otimes 1)(\underline{f} \otimes \underline{g} \otimes \underline{h}) \ . \qquad (4.29)
$$

Journal of Physics: Conference Series **634** (2015) 012004 doi:10.1088/1742-6596/634/1/012004

The pullback of this expression reads as

$$\begin{aligned} s_{\bar p}^*\big((\underline{f} \star_{F\,F'} \underline{g}\,)\star_{F\,F'} \underline{h}\big) &= \mu_{F'}\circ(\mu_{F'}\otimes\mathrm{id})\circ \mathrm{e}^{\,R}\,(\Delta\otimes\mathrm{id})F^{-1}\,(F^{-1}\otimes 1)\big(s_{\bar p}^*\underline{f}\otimes s_{\bar p}^*\underline{g}\otimes s_{\bar p}^*\underline{h}\big) \\ &= \star_{\bar p}\big(\mathrm{e}^{\,R}(s_{\bar p}^*\underline{f}\otimes s_{\bar p}^*\underline{g}\otimes s_{\bar p}^*\underline{h})\big) \end{aligned} \tag{4.30}$$

where we dropped both F^{-1} and $(\Delta\otimes\mathrm{id})F^{-1} = \exp\big(\frac12\,(\Delta(P_i)\otimes\tilde P^i - \Delta(\tilde P^i)\otimes P_i)\big)$ because they act trivially on functions of constant momentum. In particular if f,g,h are functions on configuration space and $\underline{f}=\pi^*f$, $\underline{g}=\pi^*g$, $\underline{h}=\pi^*h$, then we recover the basic triproduct $\mu_{\bar p}^{(3)}$ from (3.15), i.e.,

$$s_{\bar p}^*\big((\pi^*f \star_{F\,F'} \pi^*g\,)\star_{F\,F'} \pi^*h\big) = \star_{\bar p}\big(\mathrm{e}^{\,R}(f\otimes g\otimes h)\big) = \mu_{\bar p}^{(3)}(f\otimes g\otimes h)\ . \tag{4.31}$$

To generalise this computation to the pullback of the product of n functions we introduce the notation $F = F^\alpha\otimes F_\alpha \in U(\mathfrak{h})\otimes U(\mathfrak{h})$ (with summation over α understood) and define $F_{12} := F\otimes 1$, $F_{23} := 1\otimes F$, $F_{13} := F^\alpha\otimes 1\otimes F_\alpha$ in $U(\mathfrak{h})^{\otimes 3}$. More generally $F_{ab}\in U(\mathfrak{h})^{\otimes n}$ is the element which is non-trivial only in the a-th and b-th factors of the tensor product: $F_{ab} = 1\otimes\cdots\otimes F^\alpha\otimes\cdots\otimes F_\alpha\otimes\cdots\otimes 1$. Similarly, in $U(\mathfrak{h})^{\otimes n}$ we set $R_{123} = R\otimes 1\otimes\cdots\otimes 1$ and more generally

$$R_{abc} = \tfrac{\ell_s^4}{12}\,R^{ijk}\,P_i^a\,P_j^b\,P_k^c\ , \tag{4.32}$$

where as before $\zeta^a\in U(\mathfrak{h})^{\otimes n}$ for $\zeta\in U(\mathfrak{h})$ is the element which is non-trivial only in the a-th factor of the tensor product: $\zeta^a = 1\otimes\cdots\otimes\zeta\otimes\cdots\otimes 1$.

With these notations we have

$$\begin{aligned} (\Delta\otimes\mathrm{id})F^{-1} &= (\Delta\otimes\mathrm{id})\exp\big(\tfrac12\,(P_i\otimes\tilde P^i - \tilde P^i\otimes P_i)\big) \\ &= \exp\Big(\tfrac12\,(\Delta(P_i)\otimes\tilde P^i - \Delta(\tilde P^i)\otimes P_i)\Big) \ = \ F_{13}^{-1}\,F_{23}^{-1}\ , \end{aligned} \tag{4.33}$$

where in the last step we expanded the coproduct and then observed that the arguments of the exponential mutually commute. Substituting this equality in the second identity of Proposition 4.15 we obtain $(\Delta^{F'}\otimes\mathrm{id})F^{-1} = \mathrm{e}^{\,R}\,F_{13}^{-1}\,F_{23}^{-1}$, which is immediately generalized to

$$(\Delta^{F'}\otimes\mathrm{id}^{\otimes n-1})F_{1e}^{-1} = \mathrm{e}^{\,R_{12\,e+1}}\,F_{1\,e+1}^{-1}\,F_{2\,e+1}^{-1}\ . \tag{4.34}$$

Using as in (4.33) multiplicativity of the coproduct and commutativity of the momentum algebra we also easily obtain $(\Delta^{F'}\otimes\mathrm{id}^{\otimes 2})\,\mathrm{e}^{\,R} = \mathrm{e}^{\,R_{134}}\,\mathrm{e}^{\,R_{234}}$ which is immediately generalized to

$$(\Delta^{F'}\otimes\mathrm{id}^{\otimes n-1})\,\mathrm{e}^{\,R_{1bc}} = \mathrm{e}^{\,R_{1\,b+1\,c+1}}\,\mathrm{e}^{\,R_{2\,b+1\,c+1}}\ . \tag{4.35}$$

Proposition 4.36.

$$\begin{aligned} \big(\cdots(\underline{f}_1\star_{F\,F'}\underline{f}_2\,)\star_{F\,F'}\cdots\star_{F\,F'}\underline{f}_{n-1}\big)\star_{F\,F'}\underline{f}_n & \\ = \mu_{F'}\circ(\mu_{F'}\otimes\mathrm{id})\circ\cdots\circ(\mu_{F'}\otimes\mathrm{id}^{\otimes n-2}) & \\ \circ \prod_{1\le a<b<c\le n}\mathrm{e}^{\,R_{abc}}\ \prod_{1\le d<e\le n} F_{de}^{-1}\big(\underline{f}_1\otimes\cdots\otimes\underline{f}_n\big)\ . & \end{aligned}$$

Proof. The proof is by induction. The assertion holds for $n=3$ by (4.29). We suppose that it holds for $n>3$ and prove that it holds for $n+1$ by computing

$$\big(\cdots(\underline{f}_1\star_{F\,F'}\underline{f}_2)\star_{F\,F'}\cdots\star_{F\,F'}\underline{f}_n\big)\star_{F\,F'}\underline{f}_{n+1} \tag{4.37}$$

$$
\begin{aligned}
&= \mu_{F'} \circ (\mu_{F'} \otimes \mathrm{id}) \circ \cdots \circ (\mu_{F'} \otimes \mathrm{id}^{\otimes n-2}) \\
&\quad \circ \prod_{1\le a<b<c\le n} \mathrm{e}^{R_{abc}} \prod_{1\le d<e\le n} F_{de}^{-1} \big(\cdots (\underline{f}_1 \star_{F\,F'} \underline{f}_2) \otimes \cdots \otimes \underline{f}_n \big) \otimes \underline{f}_{n+1} \\
&= \mu_{F'} \circ (\mu_{F'} \otimes \mathrm{id}) \circ \cdots \circ (\mu_{F'} \otimes \mathrm{id}^{\otimes n-2}) \\
&\quad \circ \prod_{1\le a<b<c\le n} \mathrm{e}^{R_{abc}} \prod_{1\le d<e\le n} F_{de}^{-1} \, (\mu_{F'} \otimes \mathrm{id}^{\otimes n-1}) F_{12}^{-1} \big(\underline{f}_1 \otimes \cdots \otimes \underline{f}_n \otimes \underline{f}_{n+1} \big) \\
&= \mu_{F'} \circ (\mu_{F'} \otimes \mathrm{id}) \circ \cdots \circ (\mu_{F'} \otimes \mathrm{id}^{\otimes n-1}) \\
&\quad \circ \Big((\Delta^{F'} \otimes \mathrm{id}^{\otimes n-1}) \Big(\prod_{1\le a<b<c\le n} \mathrm{e}^{R_{abc}} \prod_{1\le d<e\le n} F_{de}^{-1} \Big) \Big) F_{12}^{-1} \big(\underline{f}_1 \otimes \cdots \otimes \underline{f}_n \otimes \underline{f}_{n+1} \big) \\
&= \mu_{F'} \circ (\mu_{F'} \otimes \mathrm{id}) \circ \cdots \circ (\mu_{F'} \otimes \mathrm{id}^{\otimes n-1}) \\
&\quad \circ \prod_{1\le a<b<c\le n+1} \mathrm{e}^{R_{abc}} \prod_{1\le d<e\le n+1} F_{de}^{-1} \big(\underline{f}_1 \otimes \cdots \otimes \underline{f}_n \otimes \underline{f}_{n+1} \big)
\end{aligned}
$$

where in the third step we used (4.20), and in the last step we used (4.35) to write

$$
(\Delta^{F'} \otimes \mathrm{id}^{\otimes n-1}) \prod_{1\le a<b<c\le n} \mathrm{e}^{R_{abc}} = \prod_{3\le b<c\le n+1} \mathrm{e}^{R_{1bc}} \prod_{2\le a<b<c\le n+1} \mathrm{e}^{R_{abc}} \tag{4.38}
$$

while from (4.34) we obtain

$$
\begin{aligned}
(\Delta^{F'} \otimes \mathrm{id}^{\otimes n-1}) \prod_{1\le d<e\le n} F_{de}^{-1} &= \prod_{3\le e<n+1} \mathrm{e}^{R_{12\,e}} \prod_{3\le e\le n+1} F_{1e}^{-1} \prod_{2\le d<e\le n+1} F_{de}^{-1} \\
&= \prod_{3\le e<n+1} \mathrm{e}^{R_{12\,e}} \prod_{1\le d<e\le n+1} F_{de}^{-1} \, F_{12}
\end{aligned} \tag{4.39}
$$

and the result follows. □

Proposition 3.9 now follows as an easy corollary. In Propositon 4.36 the inverse twists F_{de}^{-1} have been ordered on the right and therefore when we pullback this result along $s_{\bar{p}}^*$, as in (4.30) we can drop these terms because their action is trivial. We therefore obtain

$$
\begin{aligned}
&s_{\bar{p}}^* \big[\big(\cdots (\underline{f}_1 \star_{F\,F'} \underline{f}_2) \star_{F\,F'} \cdots \star_{F\,F'} \underline{f}_{n-1} \big) \star_{F\,F'} \underline{f}_n \big] \\
&\qquad = \mu_{F'} \circ (\mu_{F'} \otimes \mathrm{id}) \circ \cdots \circ (\mu_{F'} \otimes \mathrm{id}^{\otimes n-2}) \circ \prod_{1\le a<b<c\le n} \mathrm{e}^{R_{abc}} \big(\underline{f}_1 \otimes \cdots \otimes \underline{f}_n \big) \ ,
\end{aligned} \tag{4.40}
$$

and Proposition 3.9 follows immediately by setting $\underline{f}_i = \pi^* f_i$.

4.4. *Differential forms and tensor fields*

The algebraic techniques we have described in this section can be applied to any algebra carrying a Hopf algebra symmetry. Hence we can extend our results by considering the larger algebra of exterior differential forms rather than just the algebra of functions. There is a natural extension of the representation (4.2) of the Lie algebra $\mathfrak{h}$ on $C^\infty(\mathcal{M}) = \Omega^0(\mathcal{M})$ to the vector space of exterior forms $\Omega^\bullet(\mathcal{M}) = C^\infty(\mathcal{M}, \bigwedge^\bullet T^*\mathcal{M}) = \bigoplus_{n\ge 0} \Omega^n(\mathcal{M})$; it is given by the Lie derivative $\mathcal{L}$. This representation is extended to all of $U(\mathfrak{h})$ in the obvious way via $\mathcal{L}_{\xi\,\zeta} = \mathcal{L}_\xi \circ \mathcal{L}_\zeta$ for all $\xi, \zeta \in U(\mathfrak{h})$; for example $\mathcal{L}_{P_i\,P_j} = \mathcal{L}_{P_i} \circ \mathcal{L}_{P_j}$. Furthermore, $\Omega^\bullet(\mathcal{M})$ is an algebra with the associative exterior product $\wedge$, and the Hopf algebra $U(\mathfrak{h})$ is a symmetry of this algebra because of the Leibniz rule for the Lie derivative, or equivalently because the exterior product is compatible with the coproduct of elements of $U(\mathfrak{h})$ acting on forms (via the Lie derivative). For ease of notation we set $\zeta(\underline{\eta}) = \mathcal{L}_\zeta(\underline{\eta})$, for all $\zeta \in U(\mathfrak{h})$ and $\underline{\eta}\,,\, \underline{\omega} \in \Omega^\bullet(\mathcal{M})$, so that by (4.6) the compatibility condition reads as

$$\zeta(\underline{\eta} \wedge \underline{\omega}) = \wedge\big(\Delta(\zeta)(\underline{\eta} \otimes \underline{\omega})\big)\ . \tag{4.41}$$

The algebra of exterior forms $\Omega^\bullet(\mathcal{M})$ can then be deformed to the exterior algebra $\Omega^\bullet(\mathcal{M})_{\star_{F'}}$, as vector spaces $\Omega^\bullet(\mathcal{M}) = \Omega^\bullet(\mathcal{M})_{\star_{F'}}$; the deformed exterior product $\wedge_{\star_{F'}}$ in $\Omega^\bullet(\mathcal{M})_{\star_{F'}}$ is given by

$$\underline{\eta} \wedge_{\star_{F'}} \underline{\omega} = \wedge\big(F'^{-1}(\underline{\eta} \otimes \underline{\omega})\big)\ , \tag{4.42}$$

for all $\underline{\eta}\,,\, \underline{\omega} \in \Omega^\bullet(\mathcal{M})_{\star_{F'}}$.

Now all expressions in Sections 4.2 and 4.3 have been obtained by solely using:

- Hopf algebra properties of the twists F and F' of Proposition 4.15 and Corollary 4.18.
- Compatibility of the action of $U(\mathfrak{h})$ on $C^\infty(\mathcal{M})$ with the coproduct in $U(\mathfrak{h})$ and the product in $C^\infty(\mathcal{M})$ (and also in (4.24)–(4.25) the associativity of the product $\mu_{F'}$ in $C^\infty(\mathcal{M})_{\star_{F'}}$).
- Triviality of the action of momentum translations $\tilde{P}^i \in U(\mathfrak{h})$ on the images of the pullbacks $s^*_{\bar{p}}$ of sections $s_{\bar{p}} : M \to \mathcal{M}$ of constant momentum $\bar{p}$; this in particular implies that the twist F acts as the identity on the image of $s^*_{\bar{p}}$.

Since the pullbacks $s^*_{\bar{p}}$ and π^* naturally extend to exterior forms, so that the three conditions above also hold true for the algebra $\Omega^\bullet(\mathcal{M})$, we conclude that all expressions in Section 4.2 and in Section 4.3 hold true also when we replace the product μ with the exterior product $\wedge$ and functions $\underline{f}_i \in C^\infty(\mathcal{M})$ or $f_i \in C^\infty(M)$ with forms $\underline{\eta}_i \in \Omega^\bullet(\mathcal{M})$ or $\eta_i \in \Omega^\bullet(M)$, and the action of the universal enveloping algebra $U(\mathfrak{h})$ on forms is always via the Lie derivative. In particular we have

$$\begin{aligned}\wedge^{(n)}_{\bar{p}}(\eta_1 \otimes \cdots \otimes \eta_n) &:= s^*_{\bar{p}}\big[\big(\cdots(\pi^*\eta_1 \wedge_{\star_{F\,F'}} \pi^*\eta_2) \wedge_{\star_{F\,F'}} \cdots\big) \wedge_{\star_{F\,F'}} \pi^*\eta_n\big] \\ &= \wedge_{\star_{\bar{p}}} \circ \prod_{1\le a<b<c\le n} \mathrm{e}^{\,R_{abc}}\,(\eta_1 \otimes \cdots \otimes \eta_n)\ ,\end{aligned} \tag{4.43}$$

for all $\eta_i \in \Omega^\bullet(M)$, $i = 1, \ldots, n$, where $\wedge_{\star_{\bar{p}}}$ is the Moyal–Weyl star product on forms (cf. (3.8))

$$\wedge_{\star_{\bar{p}}}(\eta_1 \otimes \cdots \otimes \eta_n) = \wedge\Big[\exp\Big(\frac{\mathrm{i}\,\ell_s^4}{6\hbar} \sum_{1\le a<b\le n} \theta^{ij}_{\bar{p}}\, \mathcal{L}^a_{\partial_i}\, \mathcal{L}^b_{\partial_j}\Big)(\eta_1 \otimes \cdots \otimes \eta_n)\Big]\ . \tag{4.44}$$

In the general context of integration of forms, we also find that the products $\wedge^{(n)}_{\bar{p}}$ under integration reduce to the Moyal–Weyl exterior product $\wedge_{\star_{\bar{p}}}$.

Proposition 4.45. *If $\eta_1, \ldots, \eta_n \in \Omega^\bullet(M)$ are forms on configuration space M such that $\eta_1 \wedge \cdots \wedge \eta_n$ is a top form in $\Omega^\bullet(M)$, then*

$$\int_M \wedge^{(n)}_{\bar{p}}(\eta_1 \otimes \cdots \otimes \eta_n) = \int_M \eta_1 \wedge_{\star_{\bar{p}}} \cdots \wedge_{\star_{\bar{p}}} \eta_n\ .$$

Proof. We first compute

$$
\begin{aligned}
\wedge_{\bar{p}}^{(n)}(\eta_1\otimes\ldots\otimes\eta_n) &= \wedge_{\star_{\bar{p}}}\circ\exp\Big(\sum_{1\le a<b<c\le n} R_{abc}\Big)(\eta_1\otimes\cdots\eta_n) \qquad (4.46)\\
&= \eta_1\wedge_{\star_{\bar{p}}}\cdots\wedge_{\star_{\bar{p}}}\eta_n + \wedge_{\star_{\bar{p}}}\circ\sum_{1\le a<b<c\le n} R_{abc}\circ\mathcal{O}\,(\eta_1\otimes\cdots\otimes\eta_n)\\
&= \eta_1\wedge_{\star_{\bar{p}}}\cdots\wedge_{\star_{\bar{p}}}\eta_n\\
&\quad+\frac{\ell_s^4}{12}\wedge_{\star_{\bar{p}}}\circ R^{ijk}\sum_{1\le a<b<c\le n}\mathcal{L}^a_{\partial_i}\,\mathcal{L}^b_{\partial_j}\,\mathcal{L}^c_{\partial_k}\circ\mathcal{O}\,(\eta_1\otimes\cdots\otimes\eta_n)\\
&= \eta_1\wedge_{\star_{\bar{p}}}\cdots\wedge_{\star_{\bar{p}}}\eta_n\\
&\quad+\frac{\ell_s^4}{12}\wedge_{\star_{\bar{p}}}\circ R^{ijk}\sum_{b=2}^{n-1}\mathcal{L}^b_{\partial_j}\sum_{a=1}^{b-1}\mathcal{L}^a_{\partial_i}\sum_{c=b+1}^{n}\mathcal{L}^c_{\partial_k}\circ\mathcal{O}\,(\eta_1\otimes\cdots\otimes\eta_n)\\
&= \eta_1\wedge_{\star_{\bar{p}}}\cdots\wedge_{\star_{\bar{p}}}\eta_n\\
&\quad+\frac{\ell_s^4}{12}\wedge_{\star_{\bar{p}}}\circ R^{ijk}\sum_{b=2}^{n-1}\sum_{e=1}^{n}\mathcal{L}^e_{\partial_j}\sum_{a=1}^{b-1}\mathcal{L}^a_{\partial_i}\sum_{c=b+1}^{n}\mathcal{L}^c_{\partial_k}\circ\mathcal{O}\,(\eta_1\otimes\cdots\otimes\eta_n)\\
&= \eta_1\wedge_{\star_{\bar{p}}}\cdots\wedge_{\star_{\bar{p}}}\eta_n\\
&\quad+\frac{\ell_s^4}{12}\mathcal{L}_{\partial_j}\circ\wedge_{\star_{\bar{p}}}\circ R^{ijk}\sum_{b=2}^{n-1}\sum_{a=1}^{b-1}\mathcal{L}^a_{\partial_i}\sum_{c=b+1}^{n}\mathcal{L}^c_{\partial_k}\circ\mathcal{O}\,(\eta_1\otimes\cdots\otimes\eta_n)\ .
\end{aligned}
$$

In the second line we expanded the exponential by factoring the operator $\sum_{a<b<c} R_{abc}$ as

$$
\begin{aligned}
\exp\Big(\sum_{1\le a<b<c\le n} R_{abc}\Big) &= \mathrm{id}+\sum_{1\le a<b<c\le n} R_{abc}+\frac{1}{2}\Big(\sum_{1\le a<b<c\le n} R_{abc}\Big)^2+\cdots\\
&=: \mathrm{id}+\sum_{1\le a<b<c\le n} R_{abc}\circ\mathcal{O}\ . \qquad (4.47)
\end{aligned}
$$

In the fifth line we used antisymmetry of R^{ijk} to replace $\mathcal{L}^b_{\partial_j}$ with $\sum_{e=1}^n\mathcal{L}^e_{\partial_j}$. In the last line we used the Leibniz rule

$$
\mathcal{L}_{\partial_i}(\eta_1\wedge_{\star_{\bar{p}}}\cdots\wedge_{\star_{\bar{p}}}\eta_n)=\wedge_{\star_{\bar{p}}}\circ\sum_{e=1}^{n}\mathcal{L}^e_{\partial_j}(\eta_1\otimes\cdots\otimes\eta_n)\ . \qquad (4.48)
$$

Next we use the Cartan formula for the Lie derivative in terms of the exterior derivative and the contraction operator as $\mathcal{L}_{\partial_j}=\imath_{\partial_j}\circ\mathrm{d}+\mathrm{d}\circ\imath_{\partial_j}$, observe that when acting on a top form it simplifies to $\mathcal{L}_{\partial_j}=\mathrm{d}\circ\imath_{\partial_j}$, and the result then follows by integrating (4.46). □

Similarly to the exterior algebra one can also deform the tensor algebra. As for the deformed exterior product $\wedge_{\star_{\mathcal{F}}}$, the deformed tensor product $\otimes_{C^\infty(\mathcal{M})_{\star_{\mathcal{F}}}}$ is defined by composing the usual tensor product $\otimes_{C^\infty(\mathcal{M})}$ over $C^\infty(\mathcal{M})$ with the inverse twist: $\otimes_{C^\infty(\mathcal{M})_{\star_{\mathcal{F}}}}:=\otimes_{C^\infty(\mathcal{M})}\circ\mathcal{F}^{-1}$.

4.5. Phase space diffeomorphisms

The Drinfeld twist deformation procedure we have been implementing consists in deforming algebras that carry a compatible representation of the Hopf algebra $U(\mathfrak{h})$ (cf. the first two items in the list of Section 4.4). We recall that the representation is compatible with the product in the algebra if the action of Lie algebra elements $X \in \mathfrak{h}$ on products is given by the Leibniz rule (and hence for elements $\zeta \in U(\mathfrak{h})$ by the coproduct). We shall now apply this procedure to the Lie algebra of vector fields $\mathsf{Vect}(\mathcal{M}) = C^\infty(\mathcal{M}, T\mathcal{M})$ on phase space, i.e., the Lie algebra of infinitesimal (local) diffeomorphisms. This is a nonassociative algebra with product given by the Lie bracket $[\ \ ] : \mathsf{Vect}(\mathcal{M}) \otimes \mathsf{Vect}(\mathcal{M}) \to \mathsf{Vect}(\mathcal{M})$. Thanks to the representation (4.2) the Lie algebra $\mathfrak{h}$ can be regarded as a subalgebra of $\mathsf{Vect}(\mathcal{M})$ and therefore its action is given by the Lie derivative: $\mathcal{L}_X(\underline{u}) = [X, \underline{u}]$ for all $X \in \mathfrak{h}$, $\underline{u} \in \mathsf{Vect}(\mathcal{M})$. The compatibility condition is satisfied because the Jacobi identity $[X, [\underline{u}\,,\, \underline{v}]] = [[X, \underline{u}], \underline{v}] + [\underline{u}\,,\, [X, \underline{v}]]$ is the Leibniz rule with respect to the product $[\ \ ]$.

We can thus apply the Drinfeld twist deformation procedure with 2-cochain $\mathcal{F} = F\,F'$ and obtain the deformed algebra of vector fields $\mathsf{Vect}(\mathcal{M})_{\star_\mathcal{F}} = \mathsf{Vect}(\mathcal{M})_{\star_{F\,F'}}$, which as a vector space is the same as $\mathsf{Vect}(\mathcal{M})$ but has a deformed Lie bracket

$$\begin{aligned} [\ \ ]_{\star_\mathcal{F}} \,:\, \mathsf{Vect}(\mathcal{M})_{\star_\mathcal{F}} \otimes \mathsf{Vect}(\mathcal{M})_{\star_\mathcal{F}} &\longrightarrow \mathsf{Vect}(\mathcal{M})_{\star_\mathcal{F}} \\ \underline{u} \otimes \underline{v} &\longmapsto [\underline{u}\,,\, \underline{v}]_{\star_\mathcal{F}} := [\ \ ] \circ (F\,F')^{-1}(\underline{u} \otimes \underline{v}) \ . \end{aligned} \tag{4.49}$$

This can be realized as a deformed commutator. For this, we introduce the notation $\mathcal{F} = \mathcal{F}^\alpha \otimes \mathcal{F}_\alpha$, $\mathcal{F}^{-1} = \overline{\mathcal{F}}^\alpha \otimes \overline{\mathcal{F}}_\alpha$; define the *universal $\mathcal{R}$-matrix* $\mathcal{R} = \mathcal{F}_{21}\,\mathcal{F}^{-1}$ where $\mathcal{F}_{21} = \mathcal{F}_\alpha \otimes \mathcal{F}^\alpha$ and denote its inverse by $\mathcal{R}^{-1} = \overline{\mathcal{R}}^\alpha \otimes \overline{\mathcal{R}}_\alpha$. We then compute

$$\begin{aligned} [\underline{u}\,,\, \underline{v}]_{\star_\mathcal{F}} &= \big[\overline{\mathcal{F}}^\alpha(\underline{u}), \overline{\mathcal{F}}_\alpha(\underline{v})\big] \\ &= \overline{\mathcal{F}}^\alpha(\underline{u})\,\overline{\mathcal{F}}_\alpha(\underline{v}) - \overline{\mathcal{F}}_\alpha(\underline{v})\,\overline{\mathcal{F}}^\alpha(\underline{u}) \ = \ \underline{u} \star_\mathcal{F} \underline{v} - \overline{\mathcal{R}}^\alpha(\underline{v}) \star_\mathcal{F} \overline{\mathcal{R}}_\alpha(\underline{u}) \ , \end{aligned} \tag{4.50}$$

where we wrote the undeformed Lie bracket as a commutator and introduced the $\star_\mathcal{F}$ product between vector fields $\underline{u} \star_\mathcal{F} \underline{v} := \overline{\mathcal{F}}^\alpha(\underline{u})\,\overline{\mathcal{F}}_\alpha(\underline{v})$ which is a deformation of the product on the universal enveloping algebra of $\mathsf{Vect}(\mathcal{M})$. It is easy to see that the bracket $[\ \ ]_{\star_\mathcal{F}}$ has the $\star_\mathcal{F}$ antisymmetry property

$$[\underline{u}\,,\, \underline{v}]_{\star_\mathcal{F}} = \big[\overline{\mathcal{F}}^\alpha(\underline{u}), \overline{\mathcal{F}}_\alpha(\underline{v})\big] = -\big[\overline{\mathcal{F}}_\alpha(\underline{v}), \overline{\mathcal{F}}^\alpha(\underline{u})\big] = -\big[\overline{\mathcal{R}}^\alpha(\underline{v}), \overline{\mathcal{R}}_\alpha(\underline{u})\big]_{\star_\mathcal{F}} \ . \tag{4.51}$$

We can write this relation as $[\ \ ] \circ \mathcal{F}^{-1}(\underline{u} \otimes \underline{v}) = -[\ \ ] \circ \mathcal{F}^{-1}\mathcal{R}^{-1} \circ \sigma(\underline{u} \otimes \underline{v})$ where σ is the transposition $\sigma(\underline{u} \otimes \underline{v}) = \underline{v} \otimes \underline{u}$. This notation emphasizes the antisymmetry of the $\star_\mathcal{F}$ bracket with respect to transpositions implemented by the operator $\mathcal{R}^{-1} \circ \sigma$.

We can next implement the transposition of the elements $\underline{u}$ and $\underline{v}$ in the triple tensor product $\underline{u} \otimes (\underline{v} \otimes \underline{z}) \in \mathsf{Vect}(\mathcal{M})_{\star_\mathcal{F}} \otimes (\mathsf{Vect}(\mathcal{M})_{\star_\mathcal{F}} \otimes \mathsf{Vect}(\mathcal{M})_{\star_\mathcal{F}})$. The bracketing hints that we are later on going to apply the operator $\mathrm{id} \otimes [\ \ ]$. There is actually deeper information in the bracketing [16, 7], as it defines the action of the $\mathcal{F}$ deformed Hopf algebra of translations and Bopp shifts on triple tensor products. We therefore first have to identify $\mathsf{Vect}(\mathcal{M})_{\star_\mathcal{F}} \otimes (\mathsf{Vect}(\mathcal{M})_{\star_\mathcal{F}} \otimes \mathsf{Vect}(\mathcal{M})_{\star_\mathcal{F}})$ with $(\mathsf{Vect}(\mathcal{M})_{\star_\mathcal{F}} \otimes \mathsf{Vect}(\mathcal{M})_{\star_\mathcal{F}}) \otimes \mathsf{Vect}(\mathcal{M})_{\star_\mathcal{F}}$ as quasi-Hopf algebra representations, and this is done via the action of the inverse associator $\Phi^{-1} = \mathrm{e}^{-2R}$; then we can apply the operator $\mathcal{R}_{12}^{-1} \circ \sigma_{12} = (\mathcal{R}^{-1} \circ \sigma) \otimes \mathrm{id}$ and finally go back to $\mathsf{Vect}(\mathcal{M})_{\star_\mathcal{F}} \otimes (\mathsf{Vect}(\mathcal{M})_{\star_\mathcal{F}} \otimes \mathsf{Vect}(\mathcal{M})_{\star_\mathcal{F}})$ via e^{2R}. Thus the transposition map in this case is $\mathrm{e}^{2R} \circ \mathcal{R}_{12}^{-1} \circ \sigma_{12} \circ \mathrm{e}^{-2R}$ which, since e^{R} is central and $R_{123} = -R_{213}$, simplifies to $\mathcal{R}_{12}^{-1} \circ \sigma_{12} \circ \mathrm{e}^{-4R}$. Hence by setting (with summation over the indices $^{\bar\phi\bar\phi}$ understood)

$$\underline{u}^{\bar\phi\bar\phi} \otimes \underline{v}^{\bar\phi\bar\phi} \otimes \underline{z}^{\bar\phi\bar\phi} := \mathrm{e}^{-4R}(\underline{u} \otimes \underline{v} \otimes \underline{z}) \ , \tag{4.52}$$

it explicitly acts as

$$\begin{aligned} \mathrm{e}^{2R} \circ \mathcal{R}_{12}^{-1} \circ \sigma_{12} \circ \mathrm{e}^{-2R}(\,\underline{u} \otimes \underline{v} \otimes \underline{z}\,) &= \mathcal{R}_{12}^{-1} \circ \sigma_{12} \circ \mathrm{e}^{-4R}(\,\underline{u} \otimes \underline{v} \otimes \underline{z}\,) \\ &= \overline{\mathcal{R}}^{\alpha}(\,\underline{v}^{\bar{\phi}\bar{\phi}}\,) \otimes \overline{\mathcal{R}}_{\alpha}(\,\underline{u}^{\bar{\phi}\bar{\phi}}\,) \otimes \underline{z}^{\bar{\phi}\bar{\phi}} \,. \end{aligned} \tag{4.53}$$

Independently of its quasi-Hopf algebraic aspects the relevance of the expression (4.53) is in its appearence in the $\star_{\mathcal{F}}$ Jacobi identity.

Proposition 4.54. *The deformed Lie algebra of infinitesimal diffeomorphisms is characterized by the* $\star_{\mathcal{F}}$ *Jacobi identity*

$$\big[\,\underline{u}\,,\,[\,\underline{v}\,,\underline{z}\,]_{\star_{\mathcal{F}}}\,\big]_{\star_{\mathcal{F}}} = \big[[\,\underline{u}^{\bar{\phi}},\underline{v}^{\bar{\phi}}\,]_{\star_{\mathcal{F}}}\,,\,\underline{z}^{\bar{\phi}}\,\big]_{\star_{\mathcal{F}}} + \big[\,\overline{\mathcal{R}}^{\alpha}(\,\underline{v}^{\bar{\phi}\bar{\phi}}\,)\,,\,\big[\,\overline{\mathcal{R}}_{\alpha}(\,\underline{u}^{\bar{\phi}\bar{\phi}}\,),\underline{z}^{\bar{\phi}\bar{\phi}}\,\big]_{\star_{\mathcal{F}}}\big]_{\star_{\mathcal{F}}}\,,$$

for all $\underline{u}\,,\,\underline{v}\,,\,\underline{z} \in \mathsf{Vect}(\mathcal{M})_{\star_{\mathcal{F}}}$*, where*

$$\underline{u}^{\bar{\phi}} \otimes \underline{v}^{\bar{\phi}} \otimes \underline{z}^{\bar{\phi}} := \Phi^{-1}(\,\underline{u} \otimes \underline{v} \otimes \underline{z}\,) = \mathrm{e}^{-2R}(\,\underline{u} \otimes \underline{v} \otimes \underline{z}\,)$$

and $\underline{u}^{\bar{\phi}\bar{\phi}} \otimes \underline{v}^{\bar{\phi}\bar{\phi}} \otimes \underline{z}^{\bar{\phi}\bar{\phi}}$ *is defined in (4.52).*

Proof. We compute

$$\begin{aligned} \big[\,\underline{u}\,,\,[\,\underline{v}\,,\underline{z}\,]_{\star_{\mathcal{F}}}\,\big]_{\star_{\mathcal{F}}} &= [\quad] \circ \mathcal{F}^{-1} \circ (\mathrm{id} \otimes [\quad]) \circ (1 \otimes \mathcal{F}^{-1})(\,\underline{u} \otimes \underline{v} \otimes \underline{z}\,) \\ &= [\quad] \circ (\mathrm{id} \otimes [\quad]) \circ (\mathrm{id} \otimes \Delta)\mathcal{F}^{-1}\,(1 \otimes \mathcal{F}^{-1})(\,\underline{u} \otimes \underline{v} \otimes \underline{z}\,) \end{aligned} \tag{4.55}$$

and

$$\begin{aligned} \big[[\,\underline{u}^{\bar{\phi}},\underline{v}^{\bar{\phi}}\,]_{\star_{\mathcal{F}}}\,,\,\underline{z}^{\bar{\phi}}\,\big]_{\star_{\mathcal{F}}} &= [\quad] \circ ([\quad] \otimes \mathrm{id}) \circ (\Delta \otimes \mathrm{id})\mathcal{F}^{-1}\,(\mathcal{F}^{-1} \otimes 1)(\,\underline{u}^{\bar{\phi}} \otimes \underline{v}^{\bar{\phi}} \otimes \underline{z}^{\bar{\phi}}\,) \\ &= [\quad] \circ ([\quad] \otimes \mathrm{id}) \circ (\mathrm{id} \otimes \Delta)\mathcal{F}^{-1}\,(1 \otimes \mathcal{F}^{-1})(\,\underline{u} \otimes \underline{v} \otimes \underline{z}\,) \end{aligned} \tag{4.56}$$

where we used the 2-cochain property (4.26). It follows that

$$\begin{aligned} &\big[\,\overline{\mathcal{R}}^{\alpha}(\,\underline{v}^{\bar{\phi}\bar{\phi}}\,)\,,\,[\,\overline{\mathcal{R}}_{\alpha}(\,\underline{u}^{\bar{\phi}\bar{\phi}}\,),\underline{z}^{\bar{\phi}\bar{\phi}}\,]_{\star_{\mathcal{F}}}\big]_{\star_{\mathcal{F}}} \\ &\quad= [\quad] \circ (\mathrm{id} \otimes [\quad]) \circ (\mathrm{id} \otimes \Delta)\mathcal{F}^{-1}\,(1 \otimes \mathcal{F}^{-1})\,\mathcal{R}_{12}^{-1} \circ \sigma_{12} \circ \mathrm{e}^{-4R}(\,\underline{u} \otimes \underline{v} \otimes \underline{z}\,) \\ &\quad= [\quad] \circ (\mathrm{id} \otimes [\quad]) \circ (\Delta \otimes \mathrm{id})\mathcal{F}^{-1}\,(\mathcal{F} \otimes 1)\;\mathrm{e}^{-2R} \circ \sigma_{12} \circ \mathrm{e}^{-4R}(\,\underline{u} \otimes \underline{v} \otimes \underline{z}\,) \\ &\quad= [\quad] \circ (\mathrm{id} \otimes [\quad]) \circ \sigma_{12} \circ (\Delta \otimes \mathrm{id})\mathcal{F}^{-1}\,(\mathcal{F}^{-1} \otimes 1)\;\mathrm{e}^{-2R}(\,\underline{u} \otimes \underline{v} \otimes \underline{z}\,) \\ &\quad= [\quad] \circ (\mathrm{id} \otimes [\quad]) \circ \sigma_{12} \circ (\mathrm{id} \otimes \Delta)\mathcal{F}^{-1}\,(1 \otimes \mathcal{F}^{-1})(\,\underline{u} \otimes \underline{v} \otimes \underline{z}\,) \end{aligned} \tag{4.57}$$

where in the third line we used the 2-cochain property (4.26) and noticed that since in our case $\mathcal{F}_{21} = \mathcal{F}^{-1}$ we have $\mathcal{R}^{-1} = \mathcal{F}\,\mathcal{F}_{21}^{-1} = \mathcal{F}^2$. In the fourth line we moved the permutation σ_{12} to the left and used $\mathrm{e}^{-2R} \circ \sigma_{12} = \sigma_{12} \circ \mathrm{e}^{2R}$, $\mathcal{F} \circ \sigma = \sigma \circ \mathcal{F}^{-1}$ and the fact that the undeformed coproduct $\Delta(\zeta)$ of any element $\zeta \in U(\mathfrak{h})$ is *cocommutative*, i.e., it is symmetric with respect to exchange of its two legs, see (4.4). In the last line we again used (4.26). The proof now follows by observing that the undeformed Jacobi identity is equivalent to the equality

$$[\quad] \circ (\mathrm{id} \otimes [\quad]) = [\quad] \circ ([\quad] \otimes \mathrm{id}) + [\quad] \circ (\mathrm{id} \otimes [\quad]) \circ \sigma_{12} \tag{4.58}$$

of operators on tensor products of vector fields. □

Note that the underlying mathematical structure we are constructing is that of a quasi-Lie algebra of a quasi-Hopf algebra. The original Hopf algebra is the universal enveloping algebra of vector fields $U(\mathsf{Vect}(\mathcal{M}))$. This is deformed to a quasi-Hopf algebra $U(\mathsf{Vect}(\mathcal{M}))^{\mathcal{F}}$ via the 2-cochain $\mathcal{F} \in U(\mathfrak{h}) \otimes U(\mathfrak{h}) \subset U(\mathsf{Vect}(\mathcal{M})) \otimes U(\mathsf{Vect}(\mathcal{M}))$ (with the inclusion via the representation (4.2)). The quasi-Lie algebra of vector fields $\mathsf{Vect}(\mathcal{M})_{\star_{\mathcal{F}}}$ is the linear space $\mathsf{Vect}(\mathcal{M})$ with the bracket $[\ \ ]_{\star_{\mathcal{F}}} : \mathsf{Vect}(\mathcal{M})_{\star_{\mathcal{F}}} \otimes \mathsf{Vect}(\mathcal{M})_{\star_{\mathcal{F}}} \to \mathsf{Vect}(\mathcal{M})_{\star_{\mathcal{F}}}$.

In the commutative case the commutator $[\,\underline{u}\,,\underline{v}\,]$ is equal to the Lie derivative $\mathcal{L}_{\underline{u}}(\,\underline{v}\,)$. This motivates the definition of the $\star_{\mathcal{F}}$ Lie derivative as

$$\mathcal{L}^{\star_{\mathcal{F}}}_{\underline{u}} := \mathcal{L}_{\overline{\mathcal{F}}^{\alpha}(\,\underline{u}\,)} \circ \overline{\mathcal{F}}_{\alpha} \ , \tag{4.59}$$

for all $\underline{u}\,,\,\underline{v} \in \mathsf{Vect}(\mathcal{M})_{\star_{\mathcal{F}}}$, so that $\mathcal{L}^{\star_{\mathcal{F}}}_{\underline{u}}(\,\underline{v}\,) = [\,\underline{u}\,,\underline{v}\,]_{\star_{\mathcal{F}}}$. The $\star_{\mathcal{F}}$ Lie derivative is given by composing the usual Lie derivative with the inverse twist $\mathcal{F}^{-1}$. The definition (4.59) holds more generally when the $\star_{\mathcal{F}}$ Lie derivative acts on exterior forms or tensor fields. The same algebra as that of Proposition 4.54 shows that the $\star_{\mathcal{F}}$ Lie derivative satisfies the deformed Leibniz rule

$$\mathcal{L}^{\star_{\mathcal{F}}}_{\underline{u}}(\,\underline{\eta} \wedge_{\star_{\mathcal{F}}} \underline{\omega}\,) = \mathcal{L}^{\star_{\mathcal{F}}}_{\underline{u}^{\bar\phi}}(\,\underline{\eta}^{\bar\phi}\,) \wedge_{\star_{\mathcal{F}}} \underline{\omega}^{\bar\phi} + \overline{\mathcal{R}}^{\alpha}(\,\underline{\eta}^{\bar\phi\bar\phi}\,) \wedge_{\star_{\mathcal{F}}} \mathcal{L}^{\star_{\mathcal{F}}}_{\overline{\mathcal{R}}_{\alpha}(\,\underline{u}^{\bar\phi\bar\phi}\,)}(\,\underline{\omega}^{\bar\phi\bar\phi}\,) \ , \tag{4.60}$$

for all $\underline{u} \in \mathsf{Vect}(\mathcal{M})_{\star_{\mathcal{F}}}$, $\underline{\eta}\,,\,\underline{\omega} \in \Omega^{\bullet}(\mathcal{M})_{\star_{\mathcal{F}}}$, where $\underline{u}^{\bar\phi} \otimes \underline{\eta}^{\bar\phi} \otimes \underline{\omega}^{\bar\phi} = \mathrm{e}^{-2R}(\,\underline{u} \otimes \underline{\eta} \otimes \underline{\omega}\,)$ and $\underline{u}^{\bar\phi\bar\phi} \otimes \underline{\eta}^{\bar\phi\bar\phi} \otimes \underline{\omega}^{\bar\phi\bar\phi} = \mathrm{e}^{-4R}(\,\underline{u} \otimes \underline{\eta} \otimes \underline{\omega}\,)$. For this, we note that the Leibniz rule for the undeformed Lie derivative can be written as $\mathcal{L} \circ (\mathrm{id} \otimes \wedge) = \wedge \circ (\mathcal{L} \otimes \mathrm{id}) + \wedge \circ (\mathrm{id} \otimes \mathcal{L}) \circ \sigma_{12}$, where we used the notation $\mathcal{L}(\,\underline{u}\,,\,\underline{\eta}\,) := \mathcal{L}_{\underline{u}}(\,\underline{\eta}\,)$. The Leibniz rule for tensor fields is then obtained by replacing differential forms with tensor fields and the deformed exterior product $\wedge_{\star_{\mathcal{F}}}$ with the deformed tensor product $\otimes_{C^\infty(\mathcal{M})_{\star_{\mathcal{F}}}}$.

4.6. Configuration space diffeomorphisms

We have described the deformed Lie algebra of infinitesimal diffeomorphisms on noncommutative and nonassociative phase space, and their action on phase space forms and tensors. We next study the induced deformed Lie algebra and action in configuration space. For this, we take advantage of the global coordinate system $\{x^i, p_i\}$ on phase space to define pullbacks of vector fields (more generally we could consider any phase space $\mathcal{M}$ that is foliated with leaves isomorphic to configuration space M). Any vector field on $\mathcal{M}$ is of the form $\underline{v} = v^i(x,p)\,\partial_i + \tilde v_i(x,p)\,\tilde\partial^i$. We define the Lie subalgebra of vector fields

$${}^H\mathsf{Vect}(\mathcal{M}) = \{\underline{v} \in \mathsf{Vect}(\mathcal{M}) \mid \underline{v} = v^i(x,p)\,\partial_i\} \tag{4.61}$$

and observe that the Lie algebra $\mathfrak{h}$ leaves ${}^H\mathsf{Vect}(\mathcal{M})$ invariant: $\big[\mathfrak{h}, {}^H\mathsf{Vect}(\mathcal{M})\big] \subset {}^H\mathsf{Vect}(\mathcal{M})$, hence we can apply the deformation procedure to ${}^H\mathsf{Vect}(\mathcal{M})$ and obtain the deformed Lie algebra ${}^H\mathsf{Vect}(\mathcal{M})_{\star_{\mathcal{F}}}$.

There is an obvious one-to-one correspondence between coordinate vector fields ∂_i on configuration space M and coordinate vector fields ∂_i on phase space $\mathcal{M}$. Thus, on one hand we can now inject the Lie algebra $\mathsf{Vect}(M)$ in ${}^H\mathsf{Vect}(\mathcal{M})$ by defining

$$\pi^* : \mathsf{Vect}(M) \longrightarrow {}^H\mathsf{Vect}(\mathcal{M}) \qquad \text{with} \quad v^i\,\partial_i \longmapsto \pi^*(v^i)\,\partial_i \ . \tag{4.62}$$

On the other hand, given the section $s_{\bar p} : M \to \mathcal{M}\ ,\ x \mapsto (x,\bar p)$ we can project the Lie algebra ${}^H\mathsf{Vect}(\mathcal{M})$ to $\mathsf{Vect}(M)$ by defining

$$s^*_{\bar p} : {}^H\mathsf{Vect}(\mathcal{M}) \longrightarrow \mathsf{Vect}(M) \qquad \text{with} \quad \underline{v} = v^i\,\partial_i \longmapsto s^*_{\bar p}(v^i)\,\partial_i \ . \tag{4.63}$$

We can then characterize the deformed Lie algebra of infinitesimal diffeomorphisms $\mathsf{Vect}(M)_{\bar p}$ as the vector space $\mathsf{Vect}(M)$ with the brackets

$$[u,v]^{(2)}_{\bar p} \;:=\; s^*_{\bar p}\big([\pi^*u,\pi^*v]_{\star_{\mathcal F}}\big)\ , \tag{4.64}$$

$$\big[u\,,\,[v,z]\big]^{(3)}_{\bar p} \;:=\; s^*_{\bar p}\big(\big[\pi^*u\,,\,[\pi^*v,\pi^*z]_{\star_{\mathcal F}}\big]_{\star_{\mathcal F}}\big)\ , \tag{4.65}$$

$$\big[[u,v]\,,\,z\big]^{(3)}_{\bar p} \;:=\; s^*_{\bar p}\big(\big[[\pi^*u,\pi^*v]_{\star_{\mathcal F}}\,,\,\pi^*z\big]_{\star_{\mathcal F}}\big)\ . \tag{4.66}$$

Since no derivatives in the momentum directions appear, for two vector fields we find explicitly

$$[u,v]^{(2)}_{\bar p} = s^*_{\bar p}\circ[\ \ ]\circ F'^{-1}\,F^{-1}(\pi^*u\otimes\pi^*v) = s^*_{\bar p}\circ[\ \ ]\circ F'^{-1}(\pi^*u\otimes\pi^*v) = [u,v]_{\star_{\bar p}} \tag{4.67}$$

where the bracket

$$[u,v]_{\star_{\bar p}} := [\ \ ]\circ\exp\big(\tfrac{\mathrm{i}\,\ell_s^4}{6\hbar}\,\theta^{ij}_{\bar p}\,\mathcal L_{\partial_i}\otimes\mathcal L_{\partial_j}\big)(u\otimes v) \tag{4.68}$$

is the quantum Lie algebra bracket on Moyal–Weyl noncommutative space [4] (see [5, Section 8.2.3] for an elementary introduction). In terms of the twist $F_{\bar p}$ and its universal $\mathcal R$-matrix defined by

$$F_{\bar p} = \exp\big(-\tfrac{\mathrm{i}\,\ell_s^4}{6\hbar}\,\theta^{ij}_{\bar p}\,\mathcal L_{\partial_i}\otimes\mathcal L_{\partial_j}\big) \qquad\text{and}\qquad F^{-1}_{\bar p} = \overline F{}^{\alpha}_{\bar p}\otimes\overline F_{\bar p\,\alpha} = \exp\big(\tfrac{\mathrm{i}\,\ell_s^4}{6\hbar}\,\theta^{ij}_{\bar p}\,\mathcal L_{\partial_i}\otimes\mathcal L_{\partial_j}\big)\ ,$$

$$R_{\bar p} = F_{\bar p\,21}\,F^{-1}_{\bar p} = F^{-2}_{\bar p} \qquad\text{and}\qquad R^{-1}_{\bar p} = \overline R{}^{\alpha}_{\bar p}\otimes\overline R_{\bar p\,\alpha} = F^{2}_{\bar p}\ , \tag{4.69}$$

we have $[\ \ ]_{\star_{\bar p}} = [\ \ ]\circ F^{-1}_{\bar p}$ and the $\star_{\bar p}$ antisymmetry property

$$[u,v]_{\star_{\bar p}} = -\big[\overline R{}^{\alpha}_{\bar p}(v),\overline R_{\bar p\,\alpha}(u)\big]_{\star_{\bar p}}\ . \tag{4.70}$$

The bracket (4.67) also defines the Lie derivative on vector fields.

The explicit expression for the 3-bracket $[[u,v],z]^{(3)}_{\bar p}$ can be obtained by following the same steps as in (4.29)–(4.31). We just substitute $\mu_{F'}$ with $[\ \ ]_{F'} := [\ \ ]\circ F'^{-1}$ and functions with vector fields; we proceed similarly with the three bracket $[u,[v,z]]^{(3)}_{\bar p}$ (cf. (4.25)) and obtain

$$\begin{aligned}\big[[u,v]\,,\,z\big]^{(3)}_{\bar p} &= [\ \ ]_{\star_{\bar p}}\circ\big([\ \ ]_{\star_{\bar p}}\otimes\mathrm{id}\big)\circ\mathrm e^{R}(u\otimes v\otimes z)\ ,\\ \big[u\,,\,[v,z]\big]^{(3)}_{\bar p} &= [\ \ ]_{\star_{\bar p}}\circ\big(\mathrm{id}\otimes[\ \ ]_{\star_{\bar p}}\big)\circ\mathrm e^{-R}(u\otimes v\otimes z)\ .\end{aligned} \tag{4.71}$$

The relation between the 3-brackets $[\ [\ \ ]]^{(3)}_{\bar p}$ and $[[\ \ ]\ ]^{(3)}_{\bar p}$ can be obtained from these explicit expressions: From the $\star_{\bar p}$ antisymmetry property of the bracket $[\ \ ]_{\star_{\bar p}}$, the easily checked property of the universal $\mathcal R$-matrix $(\mathrm{id}\otimes\Delta^{F_{\bar p}})R^{-1}_{\bar p} = R^{-1}_{\bar p\,12}\,R^{-1}_{\bar p\,13}$ and the identity $\zeta\circ[\ \ ]_{\star_{\bar p}} = [\ \ ]_{\star_{\bar p}}\circ\Delta^{F_{\bar p}}(\zeta)$ (cf. (4.20)), we have

$$\big[[u,v]_{\star_{\bar p}}\,,\,z\big]_{\star_{\bar p}} = -\big[\overline R{}^{\alpha}_{\bar p}(z)\,,\,\overline R_{\bar p\,\alpha}\big([u,v]_{\star_{\bar p}}\big)\big]_{\star_{\bar p}} = -\big[\overline R{}^{\beta}_{\bar p}\,\overline R{}^{\alpha}_{\bar p}(z)\,,\,\big[\overline R_{\bar p\,\beta}(u),\overline R_{\bar p\,\alpha}(v)\big]_{\star_{\bar p}}\big]_{\star_{\bar p}}\ , \tag{4.72}$$

which implies

$$\big[[u^{\bar\phi},v^{\bar\phi}]\,,\,z^{\bar\phi}\big]^{(3)}_{\bar p} = -\big[\overline R{}^{\beta}_{\bar p}\,\overline R{}^{\alpha}_{\bar p}(z)\,,\,\big[\overline R_{\bar p\,\beta}(u),\overline R_{\bar p\,\alpha}(v)\big]\big]^{(3)}_{\bar p} \tag{4.73}$$

where we used the notation

$$u^{\bar\phi}\otimes v^{\bar\phi}\otimes z^{\bar\phi} := \mathrm e^{-2R}(u\otimes v\otimes z)\ . \tag{4.74}$$

We can now induce the deformed Jacobi identity for the 3-bracket $[\ \ [\ \ ]]^{(3)}_{\bar p}$ from the phase space $\star_{\mathcal F}$ Jacobi identity of Proposition 4.54. We let $(\underline{u}\,,\underline{v}\,,\underline{z}\,)=(\pi^*u,\pi^*v,\pi^*z)$ and then pullback along $s^*_{\bar p}$. Moving F to the right of F', so that its action becomes trivial on the image of π^*, the last term simplifies as

$$\begin{aligned}
&s^*_{\bar p}\big(\big[\overline{\mathcal R}{}^\alpha(\pi^*v^{\bar\phi\bar\phi})\,,\,\big[\overline{\mathcal R}_\alpha(\pi^*u^{\bar\phi\bar\phi}),\pi^*z^{\bar\phi\bar\phi}\big]_{\star_{\mathcal F}}\big]_{\star_{\mathcal F}}\big)\\
&=s^*_{\bar p}\big([\ \ ]_{F'}\circ(\mathrm{id}\otimes[\ \ ]_{F'})\circ(\mathrm{id}\otimes\Delta^{F'})F^{-1}\,(1\otimes F^{-1})\,(F'^{\,2}\otimes 1)\;\mathrm e^{\,4R}(\pi^*v\otimes\pi^*u\otimes\pi^*z)\big)\\
&=s^*_{\bar p}\big([\ \ ]_{F'}\circ(\mathrm{id}\otimes[\ \ ]_{F'})\circ(\Delta^{F'}\otimes\mathrm{id})\,F^{-1}\,(F'^{\,2}\otimes 1)\;\mathrm e^{\,2R}(\pi^*v\otimes\pi^*u\otimes\pi^*z)\big)\\
&=s^*_{\bar p}\big([\ \ ]_{F'}\circ(\mathrm{id}\otimes[\ \ ]_{F'})\circ\mathrm e^{\,-R}\,(F'^{\,2}\otimes 1)\;\mathrm e^{\,2R}(\pi^*v\otimes\pi^*u\otimes\pi^*z)\big)\\
&=[\ \ ]_{\star_{\bar p}}\circ(\mathrm{id}\otimes[\ \ ]_{\star_{\bar p}})\circ\mathrm e^{\,-R}\,\big(\overline R_{\bar p}{}^\alpha(v^{\bar\phi})\otimes\overline R_{\bar p\,\alpha}(u^{\bar\phi})\otimes z^{\bar\phi}\big)\\
&=\big[\overline R_{\bar p}{}^\alpha(v^{\bar\phi})\,,\,\big[\overline R_{\bar p\,\alpha}(u^{\bar\phi}),z^{\bar\phi}\big]\big]^{(3)}_{\bar p}\ .
\end{aligned}\tag{4.75}$$

In the second line we used $\mathcal R^{-1}=F'^{\,2}\,F^2$, recalled (4.52) and then the antisymmetry of the R-flux components R^{ijk}. In the third line we used Corollary 4.18, while in the fourth line we used the second identity of Proposition 4.15 and the equality $(\Delta\otimes\mathrm{id})F^{-1}\,(F'^{\,2}\otimes 1)=\mathrm e^{\,-2R}\,(F'^{\,2}\otimes 1)(\Delta\otimes\mathrm{id})F^{-1}$ that follows for example from the Baker-Campbell-Haudorff formula (4.10). In the fifth line we used again antisymmetry of the R-flux components and recalled that $R_{\bar p}^{-1}=F_{\bar p}^2$. We therefore conclude that the deformed Jacobi identity reads as

$$[u\,,\,[v,z]]^{(3)}_{\bar p}+\big[\overline R{}^\beta_{\bar p}\,\overline R{}^\alpha_{\bar p}(z)\,,\,\big[\overline R_{\bar p\,\beta}(u),\overline R_{\bar p\,\alpha}(v)\big]\big]^{(3)}_{\bar p}=\big[\overline R_{\bar p}{}^\alpha(v^{\bar\phi})\,,\,\big[\overline R_{\bar p\,\alpha}(u^{\bar\phi}),z^{\bar\phi}\big]\big]^{(3)}_{\bar p}\ .\tag{4.76}$$

These deformed Lie algebra expressions simplify considerably on the leaf with momentum $\bar p=0$, which gives the deformed Lie algebra $\mathsf{Vect}(M)_0$ of infinitesimal diffeomeorphisms on configuration space; it is characterized by an undeformed 2-bracket $[u,v]^{(2)}_0=[u,v]$, and the 3-bracket

$$[u\,,\,[v,z]]^{(3)}_0=[\ \ ]\circ(\mathrm{id}\otimes[\ \ ])\circ\mathrm e^{\,-R}(u\otimes v\otimes z)\tag{4.77}$$

that satisfies the deformed Jacobi identity

$$[u\,,\,[v,z]]^{(3)}_0+[z\,,\,[u,v]]^{(3)}_0+[v^{\bar\phi}\,,\,[z^{\bar\phi},u^{\bar\phi}]]^{(3)}_0=0\ .\tag{4.78}$$

We have derived this deformed Lie algebra structure from the one on phase space, however it is easy to verify (4.78) directly. We can also characterize the deformation via the Jacobiator

$$[u,v,z]_0:=[u\,,\,[v,z]]^{(3)}_0+[z\,,\,[u,v]]^{(3)}_0+[v\,,\,[z,u]]^{(3)}_0=[v\,,\,[z,u]]^{(3)}_0-[v^{\bar\phi}\,,\,[z^{\bar\phi},u^{\bar\phi}]]^{(3)}_0\ ,\tag{4.79}$$

i.e., $[\ \ ]_0=[\ \ ]\circ(\mathrm{id}\otimes[\ \ ])\circ\sigma_{12}\circ(\mathrm e^{\,-R}-1)$. We have thus derived a general expression to all orders in the R-flux components R^{ijk} for the Jacobiator on arbitrary vector fields; this formula nicely illustrates how nonassociativity can be explicitly manifest in a theory of gravity on configuration space.

Finally we study Lie derivatives of forms on configuration space. We define the Lie derivative $\mathcal L^{\bar p}$ by

$$\mathcal L^{\bar p}{}_u(\eta):=s^*_{\bar p}\circ\mathcal L^{\star_{\mathcal F}}_{\pi^*u}(\pi^*\eta)\ ,\tag{4.80}$$

for all $u \in \mathsf{Vect}(M)_{\bar{p}}$, $\eta \in \Omega^\bullet(M)_{\bar{p}}$. As in (4.67), since no partial derivatives in the momentum directions appear, and since $s_{\bar{p}}^* \mathrm{d}p_i = 0$, we find explicitly

$$\mathcal{L}^{\bar{p}}{}_u(\eta) = \mathcal{L}_{\overline{F}^{\alpha}_{\bar{p}}(u)}\big(\overline{F}_{\bar{p}\,\alpha}(\eta)\big) =: \mathcal{L}_u^{\star_{\bar{p}}}(\eta) \tag{4.81}$$

where we used the notation introduced in (4.69), and $\mathcal{L}^{\star_{\bar{p}}}$ is the $\star_{\bar{p}}$ Lie derivative on Moyal–Weyl noncommutative space (cf. [4, 5]). In order to write the Leibniz rule for this infinitesimal diffeomorphism we observe from (4.43) that the exterior product $\wedge_{\bar{p}}^{(2)}$ is the usual exterior product in Moyal–Weyl noncommutative space: $\wedge_{\bar{p}}^{(2)} = \wedge_{\star_{\bar{p}}}$; since the Lie derivative is also the same, we have

$$\mathcal{L}^{\bar{p}}{}_u(\eta \wedge_{\star_{\bar{p}}} \omega) = \mathcal{L}^{\bar{p}}{}_u(\eta) \wedge_{\star_{\bar{p}}} \omega + \overline{R}_{\bar{p}}{}^{\alpha}(\eta) \wedge_{\star_{\bar{p}}} \mathcal{L}^{\bar{p}}{}_{\overline{R}_{\bar{p}\,\alpha}(u)}(\omega) \ . \tag{4.82}$$

In particular if we pullback this expression to the leaf $\bar{p} = 0$ we obtain the usual undeformed Lie derivative action.

Acknowledgments
We thank R. Blumenhagen, L. Castellani, B. Jurčo, D. Lüst, D. Minic, D. Mylonas, A. Schenkel and P. Schupp for helpful discussions, and G. Amelino-Camelia, P. Martinetti and J.-C. Wallet for the invitation to submit this contribution to their proceedings. This work was initiated while R.J.S. was visiting the Department of Theoretical Physics of the University of Torino during January–February 2014, and completed while he was visiting the Alfred Renyi Institute of Mathematics in Budapest during March–April 2015; he warmly thanks both institutions for support and hospitality during his stays there. The work of R.J.S. was supported in part by the Consolidated Grant ST/L000334/1 from the UK Science and Technology Facilities Council.

[1] Aldazabal G, Marqués D and Núñez C, Double field theory: A pedagogical review, *Class. Quant. Grav.* **30** (2013) 163001 [arXiv:1305.1907 [hep-th]].
[2] Andriot D, Larfors M, Lüst D and Patalong P, (Non)commutative closed string on T-dual toroidal backgrounds, *JHEP* **1306** (2013) 021 [arXiv:1211.6437 [hep-th]].
[3] Aschieri P, Bakovič I, Jurčo B. and Schupp P, Noncommutative gerbes and deformation quantization, *J. Geom. Phys.* **60** (2010) 1754–1761 [arXiv:hep-th/0206101].
[4] Aschieri P, Dimitrijevic M, Meyer F and Wess J, Noncommutative geometry and gravity, *Class. Quant. Grav.* **23** (2006) 1883–1912 [arXiv:hep-th/0510059].
[5] Aschieri P, Dimitrijevic M, Kulish P, Lizzi F and Wess J, *Noncommutative spacetimes: Symmetries in noncommutative geometry and field theory*, Lect. Notes Phys. **774** (2009) 1–197.
[6] Bakas I and Lüst D, 3-cocycles, nonassociative star products and the magnetic paradigm of R-flux string vacua, *JHEP* **1401** (2014) 171 [arXiv:1309.3172 [hep-th]].
[7] Barnes G E , Schenkel A and Szabo R J , Nonassociative geometry in quasi-Hopf representation categories I: Bimodules and their internal homomorphisms, *J. Geom. Phys.* **89** (2015) 111–152 [arXiv:1409.6331 [math.QA]].
[8] Bergshoeffv E, Berman D S, van der Schaar J P and Sundell P, A noncommutative M-theory five-brane, *Nucl. Phys. B* **590** (2000) 173–197 [arXiv:hep-th/0005026].
[9] Berman D S, Cederwall M and Perry M J, Global aspects of double geometry, *JHEP* **1409** (2014) 066 [arXiv:1401.1311 [hep-th]].
[10] Blumenhagen R and Plauschinn E, Nonassociative gravity in string theory?, *J. Phys. A* **44** (2011) 015401 [arXiv:1010.1263 [hep-th]].
[11] Blumenhagen R, Fuchs M, Hassler F, Lüst D and Sun R, Nonassociative deformations of geometry in double field theory, *JHEP* **1404** (2014) 141 [arXiv:1312.0719 [hep-th]].
[12] Blumenhagen R, Deser A, Lüst D, Plauschinn E and Rennecke F, Non-geometric fluxes, asymmetric strings and nonassociative geometry, *J. Phys. A* **44** (2011) 385401 [arXiv:1106.0316 [hep-th]].

[13] Chu C S and Ho P M, Noncommutative open string and D-brane, *Nucl. Phys. B* **550** (1999) 151–168 [arXiv:hep-th/9812219].

[14] Condeescu C, Florakis I and Lüst D, Asymmetric orbifolds, non-geometric fluxes and noncommutativity in closed string theory, *JHEP* **1204** (2012) 121 [arXiv:1202.6366 [hep-th]].

[15] Cornalba L and Schiappa R, Nonassociative star product deformations for D-brane worldvolumes in curved backgrounds, *Commun. Math. Phys.*} **225** (2002) 33–66 [arXiv:hep-th/0101219].

[16] Drinfeld V G, *Quasi-Hopf algebras*, Leningrad Math. J. **1** (1990) 1419–1457.

[17] Freidel L, Leigh R G and Minic D, Born reciprocity in string theory and the nature of spacetime, *Phys. Lett. B* **730** (2014) 302–306 [arXiv:1307.7080 [hep-th]].

[18] Ho P M, Making nonassociative algebra associative, *JHEP* **0111** (2001) 026 [arXiv:hep-th/0103024].

[19] Ho P M and Yeh Y T, Noncommutative D-brane in nonconstant NS–NS *B*-field background, *Phys. Rev. Lett.*} **85** (2000) 5523–5526 [arXiv:hep-th/0005159].

[20] Hohm O, Lüst D and Zwiebach B, The spacetime of double field theory: Review, remarks, and outlook, *Fortsch. Phys.* **61** (2013) 926–966 [arXiv:1309.2977 [hep-th]].

[21] Hull C M, A geometry for non-geometric string backgrounds, *JHEP* **0510** (2005) 065 [arXiv:hep-th/0406102].

[22] Hull C M, Finite gauge transformations and geometry in double field theory, Preprint arXiv:1406.7794 [hep-th].

[23] Kawamoto S and Sasakura N, Open membranes in a constant *C*-field background and noncommutative boundary strings, *JHEP* **0007** (2000) 014 [arXiv:hep-th/0005123].

[24] Lüst D, T-duality and closed string noncommutative (doubled) geometry, *JHEP* **1012** (2010) 084 [arXiv:1010.1361 [hep-th]].

[25] Mylonas D and Szabo R J, Nonassociative field theory on non-geometric spaces, *Fortsch. Phys.*} **62** (2014) 727–732 [arXiv:1404.7304 [hep-th]].

[26] Mylonas D, Schupp P and Szabo R J, Membrane sigma-models and quantization of non-geometric flux backgrounds, *JHEP* **1209** (2012) 012 [arXiv:1207.0926 [hep-th]].

[27] Mylonas D, Schupp P and Szabo R J, Non-geometric fluxes, quasi-Hopf twist deformations and nonassociative quantum mechanics, *J. Math. Phys.* **55** (2014) 122301 [arXiv:1312.1621 [hep-th]].

[28] Mylonas D, Schupp P and Szabo R J, Nonassociative geometry and twist deformations in non-geometric string theory, *PoS* ICMP **2013** (2013) 007 [arXiv:1402.7306 [hep-th]].

[29] Saemann C and Szabo R J, Groupoid quantization of loop spaces, *PoS* CORFU **2011** (2011) 046 [arXiv:1203.5921 [hep-th]].

[30] Saemann C and Szabo R J, Groupoids, loop spaces and quantization of 2-plectic manifolds, *Rev. Math. Phys.*} **25** (2013) 1330005 [arXiv:1211.0395 [hep-th]].

[31] Schomerus V, D-branes and deformation quantization, *JHEP* **9906** (1999) 030 [arXiv:hep-th/9903205].

[32] Seiberg N and Witten E, String theory and noncommutative geometry, *JHEP* **9909** (1999) 032 [arXiv:hep-th/9908142].

[33] Shelton J, Taylor W and Wecht B, Non-geometric flux compactifications, *JHEP* **0510** (2005) 085 [arXiv:hep-th/0508133].

[34] Vaisman I, On the geometry of double field theory, *J. Math. Phys.* **53** (2012) 033509 [arXiv:1203.0836 [math.DG]].

CTCQG2014 IOP Publishing
Journal of Physics: Conference Series **634** (2015) 012005 doi:10.1088/1742-6596/634/1/012005

Light-like κ-deformations and scalar field theory via Drinfeld twist

Tajron Jurić[1], Stjepan Meljanac[2] and Andjelo Samsarov[3]

[1,2] Rudjer Bošković Institute, Theoretical Physics Division, Bijenička c.54, HR-10002 Zagreb, Croatia

[3] Dipartimento di Matematica e Informatica, Universita di Cagliari, viale Merello 92, 09123 Cagliari, Italy and INFN, Sezione di Cagliari

E-mail: [1] tjuric@irb.hr, [2] meljanac@irb.hr, [3] samsarov@unica.it

Abstract. In this article we will use the Drinfeld twist leading to light-like κ-deformations of Poincaré algebra. We shall apply the standard Hopf algebra methods in order to define the star-product, which shall be used to formulate a scalar field theory compatible with κ-Poincaré-Hopf algebra. Using this approach we show that there is no problem with formulating integration on κ-Minkowski space and no need for introducing a new measure. We have shown that the $\star$-product obtained from this twist enables us to define a free scalar field theory on κ-Minkowski space that is equivalent to a commutative one on a usual Minkowski space. We also discuss the interacting ϕ^4 scalar field model compatible with κ-Poincaré-Hopf algebra.

1. Introduction

Quantum field theory is a framework that describes the multi-particle systems in accordance with the quantum mechanical principles and the principles of special relativity. Therefore, the Poincaré symmetry is in the very heart of quantum field theory. It is argued that at very high energies the gravitational effects can no longer be neglected and that the spacetime is no longer a smooth manifold, but rather a fuzzy, or better to say a noncommutative space [1]. The quantum field theory on such noncommutative manifolds requires a new framework. This new framework is provided by noncommutative geometry [2]. In this framework, the search for diffeomorphisms that leave the physical action invariant leads to deformation of Poincaré symmetry, with κ-Poincaré symmetry being among the most extensively studied [3, 4, 5].

κ-deformed Poincaré symmetry is algebraically described by the κ-Poincaré-Hopf algebra and is an example of deformed relativistic symmetry that can possibly describe the physical reality at the Planck scale. κ is the deformation parameter usually interpreted as the Planck mass or some quantum gravity scale.

The deformation of the symmetry group is realized through the application of the Drinfeld twist on that group [6]. In order to implement this twisted (deformed) Poincaré symmetry into the quantum field theory, it is necessary to implement twisted statistics [7, 8], with the form of the interaction also being dictated by the quantum symmetry. The virtue of the twist formulation is that the deformed (twisted) symmetry algebra is the same as the original undeformed one and only the coalgebra structure changes, leading to the same free field structure as in the corresponding commutative field theory.

There have been attempts in the literature to obtain κ-Poincaré-Hopf algebra from the Abelian [9, 8] and Jordanian twist [10], but the problem with these twists is that they can not be expressed in terms of the Poincaré generators only. The κ-Poincaré-Hopf algebra was obtained using a twist in a Hopf algebroid approach in [11]. Particularly, a full description of deformation of Poincaré algebra in terms of the twist and the R-matrix is missing. However, for the light-like κ-deformations of Poincaré algebra there is a Drinfeld twist [11, 12, 13]. It was considered earlier in the literature [14, 15], but it was written only in the light-cone basis.

In this paper we will use the Drinfeld twist leading to light-like κ-deformations of Poincaré algebra [11, 12, 13] and apply the methods developed in [16, 17, 18, 19, 20] in order to first define the star-product and then to use it with the purpose of formulating a scalar field theory compatible with κ-Poincaré-Hopf algebra. As it turns out that the κ-Poincaré algebra acts on the κ-Minkowski space in the analogous way as the Poincaré algebra acts on the usual Minkowski space, it appears that there is no problem with formulating integration on κ-Minkowski space and no need for introducing a new measure like in [21, 22, 23, 24, 25]. The star product stemming from our Drinfeld twist has the cyclic property under the sign of integration.

The paper is organized as follows. In section 2 we outline the basic definitions of Hopf algebra, Drinfeld twist and star product. In section 3 we apply the Hopf algebra methods to Poincaré algebra and by using our light-like Drinfeld twist [11, 12, 13], derive the full κ-Poincaré-Hopf algebra and the corresponding κ-Minkowski space. After proving some useful properties of the star-product, we formulate the free scalar field theory (in section 4) and a model of interacting ϕ^4 field theory on κ-Minkowski space (in section 5). Finally, in section 6 we give our conclusion and future prospects.

2. Hopf algebra and Drinfeld twist

Let us first state some general facts about Drinfeld twists and Hopf algebras. Consider a Lie algebra $\mathfrak{g}$ and its universal enveloping algebra $\mathcal{U}(\mathfrak{g})$. $\mathcal{U}(\mathfrak{g})$ is an associative unital algebra. It is also a Hopf algebra with coproduct $\Delta : \mathcal{U}(\mathfrak{g}) \longrightarrow \mathcal{U}(\mathfrak{g}) \otimes \mathcal{U}(\mathfrak{g})$, counit $\epsilon : \mathcal{U}(\mathfrak{g}) \longrightarrow \mathbb{C}$ and antipode $S : \mathcal{U}(\mathfrak{g}) \longrightarrow \mathcal{U}(\mathfrak{g})$ satisfying

$$\begin{aligned} (\Delta \otimes 1)\Delta(\xi) &= (1 \otimes \Delta)\Delta(\xi), \\ m\left[(\epsilon \otimes 1)\Delta(\xi)\right] &= m\left[(1 \otimes \epsilon)\Delta(\xi)\right] = \xi, \\ m\left[(S \otimes 1)\Delta(\xi)\right] &= m\left[(1 \otimes S)\Delta(\xi)\right] = \epsilon(\xi)1, \end{aligned} \tag{1}$$

where $m : \mathcal{U}(\mathfrak{g}) \otimes \mathcal{U}(\mathfrak{g}) \longrightarrow \mathcal{U}(\mathfrak{g})$ is the multiplication map, 1 is the unity in $\mathcal{U}(\mathfrak{g})$ and $\xi \in \mathcal{U}(\mathfrak{g})$. For the generator $g \in \mathfrak{g}$ we have

$$\begin{aligned} \Delta(g) = g \otimes 1 + 1 \otimes g, &\quad \Delta(1) = 1 \otimes 1, \\ \epsilon(g) = 0, &\quad \epsilon(1) = 1, \\ S(g) = -g, &\quad S(1) = 1, \end{aligned} \tag{2}$$

and this extends to the whole $\mathcal{U}(\mathfrak{g})$ by requiring that Δ and ϵ are homomorphisms and linear, while S is an antihomomorphism and also linear.

The Drinfeld twist $\mathcal{F} \in \mathcal{U}(\mathfrak{g}) \otimes \mathcal{U}(\mathfrak{g})$ is an invertible element[1] of the Hopf algebra $\mathcal{U}(\mathfrak{g})$ that

[1] Here we actually extend the notion of enveloping algebra to some formal power series in $\frac{1}{\kappa}$ (we replace the field $\mathbb{C}$ with the ring $\mathbb{C}[[\frac{1}{\kappa}]]$) and we correspondingly consider the Hopf algebra $(\mathcal{U}(\mathfrak{g})[[\frac{1}{\kappa}]], \Delta, S, \epsilon, m)$, but for the sake of brevity we shall denote $\mathcal{U}(\mathfrak{g})[[\frac{1}{\kappa}]]$ by $\mathcal{U}(\mathfrak{g})$. In this context we have an additional requirement on the twist

$$\mathcal{F} = 1 \otimes 1 + \mathcal{O}(\frac{1}{\kappa}).$$

satisfies the cocycle condition

$$(\mathcal{F}\otimes 1)(\Delta\otimes 1)\mathcal{F} = (1\otimes\mathcal{F})(1\otimes\Delta)\mathcal{F} \tag{3}$$

and the normalization condition

$$m(\epsilon\otimes 1)\mathcal{F} = 1 = m(1\otimes\epsilon)\mathcal{F}. \tag{4}$$

The cocycle condition (3) is responsible for the associativity of the $\star$-product (to be defined in the next subsection). The twist operator $\mathcal{F}$ generates a new Hopf algebra $\mathcal{U}^{\mathcal{F}}(\mathfrak{g})$ given by $(\mathcal{U}^{\mathcal{F}}(\mathfrak{g}),\Delta^{\mathcal{F}},\epsilon^{\mathcal{F}},S^{\mathcal{F}},m^{\mathcal{F}})$, where $\mathcal{U}^{\mathcal{F}}(\mathfrak{g}) = \mathcal{U}(\mathfrak{g})$ as algebras and $\epsilon^{\mathcal{F}} = \epsilon$, while the new coproduct $\Delta^{\mathcal{F}}$ and antipode $S^{\mathcal{F}}$ are given by

$$\begin{aligned}\Delta^{\mathcal{F}} &= \mathcal{F}\Delta\mathcal{F}^{-1},\\ S^{\mathcal{F}} &= \chi S\chi^{-1},\end{aligned} \tag{5}$$

where $\chi^{-1} = m\left[(S\otimes 1)\mathcal{F}^{-1}\right]$.

2.1. $\star$-product
Let us consider an algebra $\mathcal{A}$ (over $\mathbb{C}[[\frac{1}{\kappa}]]$) and an action of the Lie algebra $\mathfrak{g}$ on $\mathcal{A}$, $(g,a)\mapsto g(a)\equiv g\triangleright a$ defined for all $g\in\mathfrak{g}$ and $a\in\mathcal{A}$. The action of $\mathfrak{g}$ on $\mathcal{A}$ induces an action[2] of the universal enveloping algebra $\mathcal{U}(\mathfrak{g})$ on $\mathcal{A}$. This way the algebra $\mathcal{A}$ is a $\mathcal{U}(\mathfrak{g})$-module algebra, that is, the algebra structure of the $\mathcal{U}(\mathfrak{g})$-module $\mathcal{A}$ is compatible with the $\mathcal{U}(\mathfrak{g})$ action

$$\xi\triangleright(ab) = \mu\left[\Delta(\xi)(\triangleright\otimes\triangleright)(a\otimes b)\right],\quad \xi\triangleright 1 = \epsilon(\xi)1, \tag{6}$$

for all $\xi\in\mathcal{U}(\mathfrak{g})$ and $a,b\in\mathcal{A}$, where 1 is the unit in $\mathcal{A}$, and μ is the multiplication map in the algebra $\mathcal{A}$.

For a given twist $\mathcal{F}\in\mathcal{U}(\mathfrak{g})\otimes\mathcal{U}(\mathfrak{g})$ satisfying (3) and (4) we can construct a deformed algebra $\mathcal{A}_\star$. The new algebra $\mathcal{A}_\star$ has the same vector space structure as $\mathcal{A}$, but the multiplication is changed via

$$a\star b = \mu\left[\mathcal{F}^{-1}(\triangleright\otimes\triangleright)(a\otimes b)\right]. \tag{7}$$

The new product $\star$ is called the *star*-product and it is easy to see that it is associative if $\mathcal{F}$ satisfies (3). One can show that the algebra $\mathcal{A}_\star$ is a module algebra with respect to a deformed Hopf algebra $(\mathcal{U}^{\mathcal{F}}(\mathfrak{g}),\Delta^{\mathcal{F}},\epsilon^{\mathcal{F}},S^{\mathcal{F}},m^{\mathcal{F}})$.

3. κ-Poincaré-Hopf algebra and κ-Minkowski space via Drinfeld twist

Now we will consider the Poincaré algebra $\mathfrak{p}$ as the Lie algebra $\mathfrak{g}$ and apply the formalism outlined in the previous section. The Poincaré algebra is a semidirect product of Lorentz algebra and translation algebra, and is generated by generators of boosts M_{0i}, rotations M_{ij} and translations P_μ satisfying

$$\begin{aligned}[M_{\mu\nu},M_{\lambda\rho}] &= -i(\eta_{\nu\lambda}M_{\mu\rho}-\eta_{\mu\lambda}M_{\nu\rho}-\eta_{\nu\rho}M_{\mu\lambda}+\eta_{\mu\rho}M_{\nu\lambda}),\\ [M_{\mu\nu},P_\lambda] &= -i(\eta_{\nu\lambda}P_\mu-\eta_{\mu\lambda}P_\nu).\end{aligned} \tag{8}$$

The universal enveloping algebra of Poincaré algebra $\mathcal{U}(\mathfrak{p})$ has a Hopf algebra structure. For the generators we have

$$\begin{aligned}\Delta M_{\mu\nu} = M_{\mu\nu}\otimes 1 + 1\otimes M_{\mu\nu},&\quad \Delta P_\mu = P_\mu\otimes 1 + 1\otimes P_\mu,\\ \epsilon(M_{\mu\nu}) = 0,&\quad \epsilon(P_\mu) = 0,\\ S(M_{\mu\nu}) = -M_{\mu\nu},&\quad S(P_\mu) = -P_\mu.\end{aligned} \tag{9}$$

[2] For example the element $g'g\in\mathcal{U}(\mathfrak{g})$ has the action $g'(g(a)) = g'\triangleright(g\triangleright a)$. Also note that $\triangleright:\mathcal{U}(\mathfrak{g})\otimes\mathcal{A}\longrightarrow\mathcal{A}$.

For the algebra $\mathcal{A}$ we choose the algebra of functions on Minkowski space. Thus, for any $f \in \mathcal{A}$ we have the following action $\triangleright : \mathcal{U}(\mathfrak{p}) \otimes \mathcal{A} \longrightarrow \mathcal{A}$

$$\begin{aligned} M_{\mu\nu} \triangleright f &= -i\left(x_\mu \frac{\partial f}{\partial x^\nu} - x_\nu \frac{\partial f}{\partial x^\mu}\right), \\ P_\mu \triangleright f &= -i\frac{\partial f}{\partial x^\mu}. \end{aligned} \tag{10}$$

The Drinfeld twist $\mathcal{F}$ for the light-like κ-deformation [11, 12, 13] is given by

$$\mathcal{F} = \exp\left\{ia^\alpha P^\beta \frac{\ln(1 + a \cdot P)}{a \cdot P} \otimes M_{\alpha\beta}\right\}, \tag{11}$$

where $a_\mu = \frac{1}{\kappa}u_\mu$ is the deformation vector and $u^\mu u_\mu = 0$. This is the so called light-like κ-deformation of Poincaré-Hopf algebra and it is proven in [26, 27] that $\mathcal{F}$ satisfies the conditions (3) and (4). Now, the deformed Hopf algebra $(\mathcal{U}^{\mathcal{F}}(\mathfrak{p}), \Delta^{\mathcal{F}}, \epsilon^{\mathcal{F}}, S^{\mathcal{F}}, m^{\mathcal{F}})$ for the generators is given by

$$\begin{aligned} \Delta^{\mathcal{F}} M_{\mu\nu} &= \mathcal{F}\Delta M_{\mu\nu}\mathcal{F}^{-1} = \Delta M_{\mu\nu} + (\delta^\alpha_\mu a_\nu - \delta^\alpha_\nu a_\mu)\left(P^\beta + \frac{1}{2}a^\beta P^2\right) Z \otimes M_{\alpha\beta}, \\ \Delta^{\mathcal{F}} P_\mu &= \mathcal{F}\Delta P_\mu \mathcal{F}^{-1} = \Delta P_\mu + \left[P_\mu a^\alpha - a_\mu\left(P^\alpha + \frac{1}{2}a^\alpha P^2\right) Z\right] \otimes P_\alpha, \\ S^{\mathcal{F}}(M_{\mu\nu}) &= \chi S(M_{\mu\nu})\chi^{-1} = -M_{\mu\nu} + (-a_\mu \delta^\beta_\nu + a_\nu \delta^\beta_\mu)\left(P^\alpha + \frac{1}{2}a^\alpha P^2\right) M_{\alpha\beta}, \\ S^{\mathcal{F}}(P_\mu) &= \chi S(P_\mu)\chi^{-1} = \left[-P_\mu - a_\mu\left(P_\alpha + \frac{1}{2}a_\alpha P^2\right) P^\alpha\right] Z, \\ \epsilon^{\mathcal{F}}(M_{\mu\nu}) &= 0, \quad \epsilon^{\mathcal{F}}(P_\mu) = 0, \end{aligned} \tag{12}$$

where $Z = \frac{1}{1+aP}$. This deformed Hopf algebra $(\mathcal{U}^{\mathcal{F}}(\mathfrak{p}), \Delta^{\mathcal{F}}, \epsilon^{\mathcal{F}}, S^{\mathcal{F}}, m^{\mathcal{F}})$ is exactly the κ-Poincaré-Hopf algebra $\mathcal{U}_\kappa(\mathfrak{p})$ in the natural realization [9, 28, 29] (classical basis [5, 30]).

Now we consider the algebra $\mathcal{A}_\star$ which is induced by the action of the twist (7)

$$f \star g = \mu\left[\mathcal{F}^{-1}(\triangleright \otimes \triangleright)(f \otimes g)\right], \quad f, g \in \mathcal{A}. \tag{13}$$

The algebra of functions on Minkowski space $\mathcal{A}$ is generated by commutative coordinates x_μ. Therefore, for the $\star$-product between the coordinates x_μ we have

$$\begin{aligned} x_\mu \star x_\nu &= \mu\left[\mathcal{F}^{-1}(\triangleright \otimes \triangleright)(x_\mu \otimes x_\nu)\right] \\ &= x_\mu x_\nu - ia_\nu x_\mu - i\eta_{\mu\nu}a^\alpha x_\alpha. \end{aligned} \tag{14}$$

Using (14) we can define a $\star$-commutator between coordinates

$$[x_\mu, x_\nu]_\star \equiv x_\mu \star x_\nu - x_\nu \star x_\mu = i(a_\mu x_\nu - a_\nu x_\mu). \tag{15}$$

Therefore, the noncommutative algebra $\mathcal{A}_\star$ is actually the free algebra generated by coordinates x_μ (with the $\star$-product as the multiplication in the algebra) divided by the two-sided ideal generated by (15). It is important to note that $\mathcal{A}_\star$ is isomorphic[3] as an algebra to the κ-Minkowski algebra $\hat{\mathcal{A}}$. The κ-Minkowski algebra $\hat{\mathcal{A}}$ is generated by the noncommuting coordinates $\hat{x}_\mu$ that satisfy

$$[\hat{x}_\mu, \hat{x}_\nu] = i(a_\mu \hat{x}_\nu - a_\nu \hat{x}_\mu). \tag{16}$$

[3] The isomorphism can be directly obtained by using the realization via Heisenberg algebra (for more details see the Appendix).

Journal of Physics: Conference Series **634** (2015) 012005 doi:10.1088/1742-6596/634/1/012005

4. Free scalar field theory in κ-Minkowski space

Using the twist (11) and the star product (13) we want to define an appropriate action functional $\mathcal{S}[\phi]$ for the free scalar field. But before we do that, let us analyze further the properties of the $\star$-product (13) and introduce the notion of adjoint.

The $\star$-product between plane waves is given by

$$e^{ipx} \star e^{iqx} = \mu\left[\mathcal{F}^{-1}(\triangleright \otimes \triangleright)(e^{ipx} \otimes e^{iqx})\right] = e^{i\mathcal{D}_\mu(p,q)x^\mu} \tag{17}$$

where

$$\mathcal{D}_\mu(p,q) = p_\mu(1+aq) + q_\mu - a_\mu \frac{pq}{1+ap} - \frac{1}{2}a_\mu(aq)\frac{p^2}{1+ap}, \tag{18}$$

and we used the notation $A_\mu B^\mu = AB$ and $A^\mu A_\mu = A^2$. The function $\mathcal{D}_\mu(p,q)$ defines the deformed momentum addition rule $\mathcal{D}_\mu(p,q) = p_\mu \oplus q_\mu$. It also gives the coproduct for p_μ via

$$\mathcal{D}_\mu(p \otimes 1, 1 \otimes p) = \Delta^{\mathcal{F}} p_\mu. \tag{19}$$

In what follows we shall use the results obtained so far in order to define a massive complex scalar field theory on κ-Minkowski space and to relate it to a corresponding field theory on the undeformed Minkowski space. To construct a proper noncommutative field theory, besides using a star-product, also requires an introduction of the adjoint, i.e. hermitian conjugation operation $\dagger$ (see [29] for details). The adjoint of the standard plane wave is defined by

$$(e^{ipx})^\dagger = e^{iS^{\mathcal{F}}(p)x}. \tag{20}$$

We assume that the standard commutative complex scalar field has a Fourier expansion

$$\phi(x) = \int d^n p\ \tilde{\phi}(p) e^{ipx}. \tag{21}$$

Now we can write the action for a non-interacting massive complex scalar field as

$$\begin{aligned} \mathcal{S}[\phi] &= \int d^n x\ \mathcal{L}(\phi, \partial_\mu \phi) \\ &= \int d^n x\ (\partial_\mu \phi)^\dagger \star (\partial^\mu \phi) + m^2 \int d^n x\ \phi^\dagger \star \phi. \end{aligned} \tag{22}$$

To find the relation with the corresponding commutative action on undeformed Minkowski space we shall first analyze

$$\int d^n x\ \varphi^\dagger \star \phi \tag{23}$$

for all $\varphi, \phi \in \mathcal{A}$. According to (20) and (21) the Fourier expansion of $\varphi^\dagger$ is

$$\varphi^\dagger(x) = \int d^n p\ \tilde{\varphi}^*(p) e^{iS^{\mathcal{F}}(p)x}, \tag{24}$$

where $*$ denotes the standard complex conjugation. From (17), (18), (21) and (24) we have

$$\begin{aligned} \int d^n x\ \varphi^\dagger \star \phi &= \int d^n x \int d^n p \int d^n q\ \tilde{\varphi}^*(p)\tilde{\phi}(q) e^{iS^{\mathcal{F}}(p)x} \star e^{iqx} \\ &= \int d^n p \int d^n q\ \tilde{\varphi}^*(p)\tilde{\phi}(q) \int d^n x\ e^{i\mathcal{D}_\mu(S^{\mathcal{F}}(p),q)x^\mu} \\ &= (2\pi)^n \int d^n p \int d^n q\ \tilde{\varphi}^*(p)\tilde{\phi}(q)\ \delta^{(n)}\left[\mathcal{D}\left(S^{\mathcal{F}}(p),q\right)\right]. \end{aligned} \tag{25}$$

To calculate the $\delta^{(n)}$-function we use the identity

$$\delta^{(n)}\left(F(p,q)\right) = \sum_i \frac{\delta^{(n)}(q-q_i)}{\left|\det\left(\frac{\partial F_\mu(p,q)}{\partial q_\nu}\right)\right|_{q=q_i}}. \tag{26}$$

It can be shown that [29]

$$\delta^{(n)}\left(\mathcal{D}(S^{\mathcal{F}}(p),q)\right) = \sqrt{1+a^2p^2}\;\delta^{(n)}(p-q), \tag{27}$$

where

$$\delta^{(n)}(p-q) = \frac{1}{(2\pi)^n}\int \mathrm{d}^n x\; \mathrm{e}^{i(p-q)x}. \tag{28}$$

However, since in our case we are dealing only with the light-like deformations, i.e. $a^2 = 0$, we have that

$$\int \mathrm{d}^n x\; \mathrm{e}^{iS^{\mathcal{F}}(p)x} \star \mathrm{e}^{iqx} = (2\pi)^n \delta^{(n)}(p-q). \tag{29}$$

Thus, finally[4]

$$\int \mathrm{d}^n x\; \varphi^\dagger \star \phi = \int \mathrm{d}^n x\; \varphi^* \phi. \tag{30}$$

We see that under the integration sign we can simply drop out the $\star$-product and replace it with an ordinary pointwise multiplication. This means that the free massive scalar field theory on κ-Minkowski space loses a nonlocal character and takes on an undeformed shape (at least for the mass term).

Now we have to investigate the partial integration for κ-Minkowski space using $\star$-product. By going through the same steps as in (25), that brought us to the result (30), we get the requred partial integration formula[5]

$$\int \mathrm{d}^n x\; (\partial_\mu \varphi)^\dagger \star \phi = -\int \mathrm{d}^n x\; \varphi^\dagger \star \partial_\mu \phi = -\int \mathrm{d}^n x\; \varphi^* \partial_\mu \phi. \tag{31}$$

Using the properties of the $\star$-product (30) and (31) and the expression for the action (22), we have

$$\begin{aligned}
\mathcal{S}[\phi] &= \int \mathrm{d}^n x\; (\partial_\mu \phi)^\dagger \star (\partial^\mu \phi) + m^2 \int \mathrm{d}^n x\; \phi^\dagger \star \phi \\
&= -\int \mathrm{d}^n x\; \phi^* \partial_\mu \partial^\mu \phi + m^2 \int \mathrm{d}^n x\; \phi^* \phi \\
&= \int \mathrm{d}^n x\; \left[(\partial_\mu \phi)^* (\partial^\mu \phi) + m^2 \phi^* \phi\right].
\end{aligned} \tag{32}$$

Eq. (32) illustrates the equivalence of the noncommutative free field theory with the commutative one. Action (32) is Poincaré invariant and it should be noted that the Klein-Gordon operator is undeformed, i.e. $-\partial_\mu \partial^\mu + m^2$.

[4] For the general deformation a_μ we would have $\int \mathrm{d}^n x\; \varphi^\dagger \star \phi = \int \mathrm{d}^n x\; \varphi^* \sqrt{1-a^2\partial^2}\phi$ (for more details see [29]).

[5] In the case that the derivative is not affected by the adjoint action we would have

$$\int \mathrm{d}^n x\; \partial_\mu \varphi^\dagger \star \phi = \int \mathrm{d}^n x\; \varphi^\dagger \star S^{\mathcal{F}}(\partial_\mu)\phi = \int \mathrm{d}^n x\; \varphi^* S^{\mathcal{F}}(\partial_\mu)\phi,$$

which then leads to a deformed Klein-Gordon operator that is not the Casimir operator of κ-Poincaré algebra [29].

5. The ϕ^4 scalar model on κ-Minkowski space

In the previous section we have investigated the free scalar field in the κ-Minkowski space and now we want to introduce the interaction between them. In order to introduce the self-interacting ϕ^4 terms into the action functional, we should replace the pointwise multiplication with the $\star$-product. This means that we should incorporate six terms corresponding to all possible permutations of fields ϕ and $\phi^\dagger$, but due to the properties of the $\star$-product under the integration sign, (30), these six permutations can be reduced to just two mutually nonequivalent terms and we can write the action for the noncommutative ϕ^4 model as

$$\begin{aligned} \mathcal{S}[\phi] = &\int \mathrm{d}^n x \, (\partial_\mu\phi)^\dagger \star (\partial^\mu\phi) + m^2 \int \mathrm{d}^n x \, \phi^\dagger \star \phi \\ &+ \frac{\lambda}{8}\int \mathrm{d}^n x \left(\phi^\dagger \star \phi^\dagger \star \phi \star \phi + \phi^\dagger \star \phi \star \phi^\dagger \star \phi\right). \end{aligned} \tag{33}$$

The first part of the action reduces to the free commutative situation (32), while the interaction part of the action is very complicated and here we use (13), (17), (18), (30) and (31) in order to expand the interaction part in the deformation parameter $\frac{1}{\kappa}$ and to further simplify the expression for the action functional. Finally we get

$$\begin{aligned} \mathcal{S}[\phi] = &\int \mathrm{d}^n x \left[(\partial_\mu\phi)^*(\partial^\mu\phi) + m^2\phi^*\phi + \frac{\lambda}{4}(\phi^*\phi)^2\right] \\ &+ i\frac{\lambda}{4}\int \mathrm{d}^n x \Big[a_\mu x^\mu \left((\phi^*)^2(\partial_\nu\phi)\partial^\nu\phi - \phi^2(\partial_\nu\phi^*)\partial^\nu\phi^*\right) \\ &\quad + a_\nu x^\mu \left(\phi^2(\partial_\mu\phi^*)\partial^\nu\phi^* - (\phi^*)^2(\partial_\mu\phi)\partial^\nu\phi\right) \\ &\quad + \frac{1}{2}a_\nu x^\mu \phi^*\phi \left((\partial_\mu\phi^\dagger)\partial^\nu\phi - (\partial_\mu)\partial^\nu\phi^*\right)\Big] + \mathcal{O}\left(\frac{1}{\kappa^2}\right). \end{aligned} \tag{34}$$

The action obtained in (34) via twist $\mathcal{F}$ (11) is exactly the same as in [31] where the theory of realizations was used [28, 9, 29].

In [31] it is argued that the changes in the particle statistics (due to twist $\mathcal{F}$) are encoded in the nonabelian momentum addition law [32, 33, 34, 35]. It can be further shown that the accordingly induced deformation of δ-function yields the usual δ-function multiplied by a certain statistical factor. Thus, in order to obtain Feynman rules in the momentum space one can use the following two steps:

(i) Use the standard commutative methods of QFT in obtaining the propagators and vertex by treating the noncommutative contribution to the action (34) just as a small perturbation.

(ii) Include the twisted nature of particle statistics. Namely, the statistics of particles is twisted and this can be implemented by requiring that the ordinary addition/subtraction rules for the particle momentum are deformed. Particularly, this means that we have

$$\sum_i k_i^\mu \longrightarrow \sum_{\oplus i} k_i^\mu \quad \& \quad \sum_i k_i^\mu - \sum_j p_j^\mu \longrightarrow \sum_{\oplus i} k_i^\mu \ominus \sum_{\oplus j} p_j^\mu \tag{35}$$

where $p_\mu \oplus q_\mu \equiv \mathcal{D}_\mu(p,q)$ and $p_\mu \ominus q_\mu \equiv p_\mu \oplus S_\mu^{\mathcal{F}}(q)$. Here $D_\mu(p,q)$ is given by (18) and note that due to the relation that exists between the coproduct and antipode we have $p_\mu \oplus S_\mu^{\mathcal{F}}(p) = 0$ (this is induced by the underlining Hopf algebra structure).

Using the above explained idea, the two-point connected Green's function up to one loop corrections takes the following form

$$G(k_1,k_2) = (2\pi)^n\delta^{(n)}(k_1-k_2)\frac{i}{k_1^2+m^2-\Pi(k_1)}, \tag{36}$$

where

$$\Pi(k_1) = \frac{\lambda m^2}{32\pi^2}\left[(1+3ak_1)\left(\frac{2}{\epsilon}+\psi(2)+\ln\frac{4\pi\mu^2}{m^2}\right)-\frac{9}{4}ak_1\right] \tag{37}$$

is the tadpole diagram contribution to the self-energy of the scalar particle [31]. Here we used the Euclidean prescription and the $n = 4-\epsilon$ dimensional regularization[6]. After introducing the appropriate counterterm, we get the finite two-point function

$$G_{fin}(k_1,k_2) = \frac{G}{1+G\frac{\lambda m^2}{32\pi^2}\left[f-(1+3ak_1)\left(\frac{2}{\epsilon}+\psi(2)+\ln\frac{4\pi\mu^2}{m^2}-\frac{9}{4}\frac{ak_1}{1+3ak_1}\right)\right]}, \tag{38}$$

where

$$G = \delta^{(n)}(k_1-k_2)\frac{i}{k_1^2+m^2} \tag{39}$$

and f is an arbitrary dimensionless function fixed by the renormalization condition.

6. Final remarks

The integral measure problems on κ-Minkowski space are avoided by using the Drinfeld twist operator (11) for the light-like κ-deformation of Poincaré-Hopf algebra. We have shown that the $\star$-product obtained from the twist (11) has very nice properties under the integration sign and enables us to define a free scalar field theory on κ-Minkowski space that is equivalent to a commutative one on a usual Minkowski space.

We present the first interacting scalar field theory compatible with κ-Poincaré-Hopf algebra obtained via Drinfeld twist (11). The action functional (33) defines a ϕ^4 scalar model on κ-Minkowski space. The truncated κ-deformed action (34) does not have the UV/IR mixing [36].

In the computation of the two-point function we have implemented the *hybrid* approach [31], i.e. we have used the deformed momentum addition/subtraction (35). There are new non-vanishing contributions even in $n = 4$ dimensions. They are arising from the κ-deformed momentum addition/subtraction and the corresponding energy-momentum conservation rule stemming from the deformed δ-function [31]. For the conserved external momenta, i.e. $k_1 = k_2$ we obtained the two-point function (36), where the finite parts represent the modification of the scalar field self-energy Π (37) and depend explicitly on the regularization parameter μ^2 and the mass of the scalar field m^2. It is important to note that (37) contains the finite correction ak_1 due to the deformed statistics on the κ-Minkowski space, thus, we obtained the explicit dependance on the direction of the propagation energy, its scale $|k_1| = E$ and the deformation parameter $\frac{1}{\kappa}$.

In this paper we have used the truncated action (34), but if we could find the $\star$-product (13) in closed form, we could express the interacting part of the action (33) up to all orders in $\frac{1}{\kappa}$ and try to obtain a one loop correction to the two-point function valid in all orders in $\frac{1}{\kappa}$. So far we have analyzed the *hybrid* version of the ϕ^4 scalar field theory. The next step would be to investigate the interacting quantum field theory on κ-Minkowski space, but with the proper notion of twisted particle statistics. In order to do this , we have to investigate the R-matrix [37] which would modify the quantization procedure, that is we have to modify the algebra of creation and annihilation operators via

$$\phi(x)\otimes\phi(y) - R\phi(y)\otimes\phi(x) = 0 \tag{40}$$

which would modify the usual spin-statistics relations of the free bosons at Planck scale. The R-matrix is defined by the twist operator $R = \tilde{\mathcal{F}}\mathcal{F}^{-1}$. The idea is that the R-matrix would enable us

[6] For more details see [31].

to define particle statistics and to properly quantize fields, while the twist operator would provide the star-product, which is crucial for writing the action in terms of commutative variables. In the end, this would enable us to derive the Feynman rules and find the noncommutative correction to the propagator and the vertex to all orders in $\frac{1}{\kappa}$.

In order to investigate gauge field theories on κ-Minkowski space, the notion of noncommutative differential geometry is necessary. The classification of all bicovariant differential calculi on κ-Minkowski space and the concept of classical noncommutative field theory was developed in [13]. This was obtained by embedding the algebra of coordinates and one-forms into a super-Heisenberg algebra. It will be very interesting to apply the Drinfeld twist (11) for the light-like κ-Poincaré algebra to obtain all aspects of noncommutative differential geometry and to formulate a noncommutative gauge field theory.

Acknowledgments
The work by T. J. and S. M. has been fully supported by Croatian Science Foundation under the project (IP-2014-09-9582). The work by A.S. was supported by the European Commission and the Croatian Ministry of Science, Education and Sports through grant project financed under the Marie Curie FP7-PEOPLE-2011-COFUND, project NEWFELPRO.

Appendix A. Realization via Heisenberg algebra

The realization of the κ-Minkowski space is defined by the unique map from the algebra $\mathcal{A}$ of functions f in commuting coordinates x_μ to the algebra $\hat{\mathcal{A}}$ of noncommuting "functions" $\hat{f}$ generated by noncommutative coordinates $\hat{x}_\mu$. This map $\Omega : \mathcal{A} \longrightarrow \hat{\mathcal{A}}$ is uniquely characterized by $f \mapsto \hat{f}$ such that

$$\hat{f} \triangleright 1 = f. \tag{A.1}$$

The action $\triangleright : \mathcal{H} \otimes \mathcal{A} \longrightarrow \mathcal{A}$ is defined by

$$p_\mu \triangleright 1 = 0, \quad x_\mu \triangleright 1 = x_\mu, \quad p_\mu \triangleright x_\nu = -i\eta_{\mu\nu}1, \tag{A.2}$$

and $\mathcal{H}$ is the Heisenberg algebra generated by x_μ and p_μ such that

$$[x_\mu, x_\nu] = [p_\mu, p_\nu] = 0, \quad [p_\mu, x_\nu] = -i\eta_{\mu\nu}1. \tag{A.3}$$

We can find the explicit realization of noncommuting coordinates $\hat{x}_\mu$ and Poincaré generators $M_{\mu\nu}$ and P_μ via generators of $\mathcal{H}$

$$\begin{aligned} \hat{x}_\mu &= x_\mu + a^\alpha M_{\mu\alpha}, \\ M_{\mu\nu} &= x_\mu p_\nu - x_\nu p_\mu, \\ P_\mu &= p_\mu. \end{aligned} \tag{A.4}$$

The star product is given by

$$f \star g = \hat{f}\hat{g} \triangleright 1 = \hat{f} \triangleright \hat{g}, \tag{A.5}$$

where $\hat{f} \triangleright 1 = f$ and $\hat{g} \triangleright 1 = g$. This gives the isomorphism $\hat{\mathcal{A}} \cong \mathcal{A}_\star$.

[1] S. Doplicher, K. Fredenhagen, J. E. Roberts, Spacetime Quantization Induced by Classical Gravity,Phys. Lett. B 331 (1994) 39;
S. Doplicher, K. Fredenhagen, J. E. Roberts,The quantum structure of spacetime at the Planck scale and quantum fields Commun. Math. Phys. 172 (1995) 187
[2] A. Connes, Noncommutative Geometry , Academic Press, 1994
[3] Lukierski J and Ruegg H, Quantum kappa Poincare in any dimension, Phys. Lett. B **329**, 189 (1994) *Preprint* [hep-th/9310117]

[4] Majid S and Ruegg H, Bicrossproduct structure of kappa Poincare group and noncommutative geometry, Phys. Lett. B **334**, 348 (1994) *Preprint* [hep-th/9405107]

[5] Kowalski-Glikman J and Nowak S, Doubly special relativity theories as different bases of kappa Poincare algebra, Phys. Lett. B **539**, 126 (2002).

[6] Chaichian M, Kulish P P, Nishijima K and Tureanu A, On a Lorentz-invariant interpretation of noncommutative space-time and its implications on noncommutative QFT, Phys. Lett. B **604**, 98 (2004).

[7] Balachandran A P, Govindarajan T R, Mangano G, Pinzul A, Qureshi B A and Vaidya S, Statistics and UV-IR mixing with twisted Poincare invariance, Phys. Rev. D **75**, 045009 (2007) *Preprint* [hep-th/0608179]

[8] Govindarajan T R, Gupta K S, Harikumar E, Meljanac S and Meljanac D, Twisted statistics in kappa-Minkowski spacetime, Phys. Rev. D **77**, 105010 (2008) *Preprint* [arXiv:0802.1576 [hep-th]]

[9] Meljanac S, Samsarov A, Stojic M and Gupta K S Kappa-Minkowski space-time and the star product realizations, Eur. Phys. J. C **53**, 295 (2008) *Preprint* [arXiv:0705.2471 [hep-th]]

[10] Borowiec A and Pachol A, kappa-Minkowski spacetime as the result of Jordanian twist deformation, Phys. Rev. D **79**, 045012 (2009) *Preprint* [arXiv:0812.0576 [math-ph]]

[11] Juric T, Kovacevic D and Meljanac S, κ-Deformed Phase Space, Hopf Algebroid and Twisting, SIGMA **10**, 106 (2014) *Preprint* [arXiv:1402.0397 [math-ph]]

[12] Juric T, Meljanac S and Strajn R, Universal κ-Poincar covariant differential calculus over κ-Minkowski space, Int. J. Mod. Phys. A **29**, 1450121 (2014) *Preprint* [arXiv:1312.2751 [hep-th]]

[13] Juric T, Meljanac S, Pikutic D and Strajn R, Toward the classification of differential calculi on κ-Minkowski space and related field theories, *Prepint* arXiv:1502.02972 [hep-th]

[14] Kulish P P, Lyakhovsky V D and Mudrov A I, Extended jordanian twists for Lie algebras, J. Math. Phys. **40**, 4569 (1999) *Preprint* [math/9806014 [math.QA]]

[15] Borowiec A and Pachol A, Unified description for κ-deformations of orthogonal groups, Eur. Phys. J. C **74**, no. 3, 2812 (2014) *Preprint* [arXiv:1311.4499 [math-ph]]

[16] V. G. Drinfeld, On constant quasiclassical solutions of the Yang-Baxter equations, Soviet Math. Dokl. 28 (1983) 667-671

[17] V. G. Drinfeld, Quasi-Hopf Algebras, Leningrad Math. J. 1 (1990) 1419 [Alg. Anal. 1N6 (1989) 114]

[18] Aschieri P, Lizzi F and Vitale P, Twisting all the way: From Classical Mechanics to Quantum Fields, Phys. Rev. D **77**, 025037 (2008) *Preprint* [arXiv:0708.3002 [hep-th]]

[19] Aschieri P, Lectures on Hopf algebras, quantum groups and twists, proceedings of the Second Modave Summer School In Mathematical Physics, International Solvay Intitutes, 208-219, *Preprint* [arXiv:hep-th/0703013] Aschieri P, Dimitrijevic M, Kulish P, Lizzi F, Wess J, Noncommutative Spacetimes, Lecture Notes in Physics , Vol. 774 ISBN: 978-3-540-89792-7,2009, Springer , Chapter 7

[20] Aschieri P, "Star Product Geometries," Russ. J. Math. Phys. **16**, 371 (2009) *Preprint* [arXiv:0903.2457 [math.QA]]

[21] Felder G and Shoikhet B, Deformation quantization with traces, *Preprint* math/0002057 [math-qa]

[22] Dimitrijevic M, Jonke L, Moller L, Tsouchnika E, Wess J and Wohlgenannt M, Deformed field theory on kappa space-time, Eur. Phys. J. C **31**, 129 (2003) *Preprint* [hep-th/0307149]

[23] Dimitrijevic M, Meyer F, Moller L and Wess J, Gauge theories on the kappa Minkowski space-time, Eur. Phys. J. C **36**, 117 (2004) *Preprint*[hep-th/0310116]

[24] Moller L, A Symmetry invariant integral on kappa-deformed spacetime, *J. High Energy Phys.* JHEP **0512**, 029 (2005) *Preprint* [hep-th/0409128]

[25] Agostini A, Amelino-Camelia G, Arzano M and D'Andrea F, A cyclic integral on kappa-Minkowski noncommutative space-time, Int. J. Mod. Phys. A **21**, 3133 (2006)

[26] Meljanac S and Pikutic D, Light-like kappa-Poincare/conformal Hopf algebra and covariant Drinfeld twist, Proceeding of XXXIII M. Born Symposium, June 2014, Wroclaw, Poland

[27] Juric T, Meljanac S, Pikutic D, Realizations of κ-Minkowski space, Drinfeld twists and twisted symmetry algebras, [in preparation]

[28] Govindarajan T R, Gupta K S , Harikumar E, Meljanac S and Meljanac D, Deformed Oscillator Algebras and QFT in kappa-Minkowski Spacetime, Phys. Rev. D **80**, 025014 (2009) *Preprint* [arXiv:0903.2355 [hep-th]] Kovacevic D and Meljanac S, Kappa-Minkowski spacetime, Kappa-Poincare Hopf algebra and realizations, J. Phys. A **45**, 135208 (2012) *Preprint* [arXiv:1110.0944 [math-ph]]

[29] Meljanac S and Samsarov A, Scalar field theory on kappa-Minkowski spacetime and translation and Lorentz invariance, Int. J. Mod. Phys. A **26**, 1439 (2011) *Preprimt* [arXiv:1007.3943 [hep-th]]

[30] Borowiec A and Pachol A, Classical basis for kappa-Poincare algebra and doubly special relativity theories, J. Phys. A **43**, 045203 (2010) *Preprint* [arXiv:0903.5251 [hep-th]]

[31] Meljanac S, Samsarov A, Trampetic J and Wohlgenannt M, Scalar field propagation in the phi^4 kappa-Minkowski model, *J. High Energy Phys.* JHEP **1112**, 010 (2011) *Preprint* [arXiv:1111.5553 [hep-th]]

[32] Freidel L, Kowalski-Glikman J and Nowak S, From noncommutative kappa-Minkowski to Minkowski space-

time, Phys. Lett. B **648**, 70 (2007) *Preprint* [hep-th/0612170]

[33] Kowalski-Glikman J and Walkus A, Star product and interacting fields on kappa-Minkowski space, Mod. Phys. Lett. A **24**, 2243 (2009) *Preprint* [arXiv:0904.4036 [hep-th]]

[34] Kowalski-Glikman J, De sitter space as an arena for doubly special relativity, Phys. Lett. B **547**, 291 (2002) *Preprint* [hep-th/0207279]

[35] Lukierski J and Nowicki A, Nonlinear and quantum origin of doubly infinite family of modified addition laws for four momenta, Czech. J. Phys. **52**, 1261 (2002) *Preprint* [hep-th/0209017]

[36] Grosse H and Wohlgenannt M, On kappa-deformation and UV/IR mixing, Nucl. Phys. B **748**, 473 (2006) *Preprint* [hep-th/0507030]

[37] Meljanac S, Samsarov A and Strajn R, Kappa-deformation of Heisenberg algebra and coalgebra: generalized Poincare algebras and R-matrix, *J. High Energy Phys.* JHEP **1208**, 127 (2012) *Preprint* [arXiv:1204.4324 [math-ph]]

Journal of Physics: Conference Series **634** (2015) 012006 doi:10.1088/1742-6596/634/1/012006

Spectral theorem in noncommutative field theories: Jacobi dynamics

Antoine Géré[1] **and Jean-Christophe Wallet**[2]

[1]Dipartimento di Matematica, Università di Genova, Via Dodecaneso, 35, I-16146 Genova, Italy

[2]Laboratoire de Physique Théorique d'Orsay, Bât. 210, CNRS and Université Paris-Sud 11, 91405 Orsay Cedex, France

E-mail: gere@dima.unige.it, jean-christophe.wallet@th.u-psud.fr

Abstract. Jacobi operators appear as kinetic operators of several classes of noncommutative field theories (NCFT) considered recently. This paper deals with the case of bounded Jacobi operators. A set of tools mainly issued from operator and spectral theory is given in a way applicable to the study of NCFT. As an illustration, this is applied to a gauge-fixed version of the induced gauge theory on the Moyal plane expanded around a symmetric vacuum. The characterization of the spectrum of the kinetic operator is given, showing a behavior somewhat similar to a massless theory. An attempt to characterize the noncommutative geometry related to the gauge fixed action is presented. Using a Dirac operator obtained from the kinetic operator, it is shown that one can construct an even, regular, weakly real spectral triple. This spectral triple does not define a noncommutative metric space for the Connes spectral distance.

1. Introduction

Noncommutative Geometry (NCG) [1] (see also [2]) may provide a way to escape physical obstructions to the existence of continuous space-time and commuting coordinates at the Planck scale [3]. This has reinforced the interest in noncommutative field theories (NCFT), which appeared in their modern formulation first in String field theory [4], followed by models on the fuzzy sphere and almost commutative geometries [5], [6] while NCFT on noncommutative Moyal spaces received attention from the end of the 90's. For reviews, see for instance [7].

It appears that the kinetic operators for the actions of several classes of NCFT considered so far are Jacobi operators [8]. Jacobi operators can be presented informally as symmetric tridiagonal infinite matrices related to a 3-term recurrence (defining) relation [8]. It turns out that Jacobi kinetic operators manifest themselves under this matrix form when the action is expressed within a so called matrix base formalism. Of course, not all NCFT have Jacobi kinetic operators but this latter case encompasses several NCFT of interest. This is for instance the case of the Grosse-Wulkenhaar model on $\mathbb{R}^4_\theta$ and $\mathbb{R}^2_\theta$ [9], [10], its rotationally invariant descendants [11] and fermionic extensions [12], the scalar ϕ^4 model on $\mathbb{R}^3_\lambda$ [13] together with the massless gauge theory on $\mathbb{R}^3_\lambda$ [14] whose truncation to the fuzzy sphere is related to the Alekseev-Recknagel- Schomerus matrix model [15].

To a given Jacobi operator corresponds a given family of orthogonal polynomials. This is the actual reason why definite families of orthogonal polynomials are automatically singled out in the diagonalization procedure of the kinetic operators for the above mentionned NCFT, and exemplified in this paper for another specific NCFT. Orthogonal polynomials are subject of intense activity in applied mathematics, ranking from approximation theory and numerical calculus to constructions of new families of multivariable orthogonal polynomials and families of random variables. The NCFT considered in this paper corresponds to Chebyshev polynomial of 2d kind, a particular type of orthogonal Jacobi polynomials. These determine entirely the kinetic operator via the spectral theorem, hence the (free) dynamics of the system. This explains the statement "Jacobi dynamics" in the title of the paper.

The numerous families of orthogonal polynomials are classified within the Askey scheme that will not be essential in the present paper. For more details, see e.g [16] and references therein. The fact that an orthogonality relation can be defined within the above orthogonal families stems from the existence of an integration measure which is a mere consequence of the spectral theorem, sometimes put in a particular form known as the Favard theorem [17] which may be traced back to works of Stieltjes [18]. The measure can be determined from its moments, giving rise to the so called (determinate or indeterminate) moment problem, that will not be central in the present paper. Note that the notion of moments is familiar to the theory of probability. There is a huge mathematical literature related to these topics, among which the reader is invited to see e.g [8], [19], [20], [21].

In this paper, we consider the case of bounded Jacobi operators. We select among the different mathematical domains mentionned above the mathematical tools relevant to the construction and the study of (quantum) NCFT, focusing on functional analysis formalism which is the most suitable one to deal with theories expressed within "matrix bases". The corresponding mathematical toolkit tailored for a use in NCFT is presented in the Section 2. Note that many of the properties given in Section 2 still apply to the case of unbounded Jacobi operators. This latter case will be presented elsewhere [22].

As a representative example, we apply the material of Section 2 to a NCFT on the Moyal plane $\mathbb{R}^2_\theta$. This is presented in the Section 3. The NCFT is a gauge-fixed version of the so called "induced gauge theory action" [23], [24] (see (3.1) in Section 3), expanded around a particular symmetric vacuum [25] which has been considered recently [26]. The NCFT has a bounded Jacobi kinetic operator and polynomial interactions after suitable gauge fixing. A complete characterization of the spectrum of the kinetic operator is given together with a very simple derivation of the propagator, stemming directly from the spectral theorem. A review on the situation of noncommutative gauge theories is given in subsection 3.1.

The induced gauge theory action (3.1) can be related to a particular NCG described by a finite volume spectral triple [27], homothetic as a noncommutative metric space [28] to the standard noncommutative Moyal space [29] whose Connes spectral distance has been investigated in [30]. Note that a theory of quantum (i.e noncommutative) compact metric spaces has been set up in [31]. An extension to the (noncommutative analog of) locally compact case has been given in [32], to which Moyal spaces belong. However, the NCFT of section 3 results from (3.1) after an expansion around some vacuum and a gauge-fixing which thus modify the initial kinetic operator in (3.1) and so the related Dirac operator. One question would be to characterize the NCG related to the resulting NCFT which would require to built a spectral triple and then perform the full computation of some spectral action with relevant perturbation of the Dirac operator. The Section 4 is a partial attempt in that direction. We construct spectral triples from a Dirac operator whose square equals the bounded Jacobi kinetic operator. In the section 5, we discuss the results and conclude.

We find a complete characterization of the spectrum of the kinetic operator. In particular, its spectrum involves 0 as a limit point which may be interpreted as a massless kinetic operator. The corresponding unbounded propagator is obtained as a simple application of Section 2, therefore completing the first analysis performed in [26]. At the perturbative level, the NCFT exhibits unsuppressed correlations at large separation between indices. By using a Dirac operator built from the kinetic operator, we show that one can construct a spectral triple. This spectral triple cannot defines a noncommutative metric space for the Connes spectral distance, i.e is not a spectral metric space in the sense of [28], [33].

2. Toolkit for NCFT: Spectral theorem and Jacobi operators

This section is written for physicists. We collect the necessary (basic) mathematical background needed to characterize spectral properties underlying the actions of several classes of NCFT as well as matrix models recently considered. The material is based on Jacobi operators, orthogonal polynomials together with a special version of the spectral theorem, sometimes called the Favard theorem. General definitions and usefull properties of operator algebra are collected in Appendix A to make the presentation self contained. In particular, we explicitely state in Appendix A the special version of the spectral theorem that is at the crossroad of the Jacobi operators, orthogonal polynomials used here and spectral properties of kinetic operators of the NCFT considered in the paper. For more details, the reader should borrow material from e.g [8], [19], [20], [21].

Let $\ell^2(\mathbb{N})$ be the Hilbert spaces of square integrable sequences $(u_n)_{n\in\mathbb{N}}$ with canonical Hilbert product $\langle (u_n),(v_m)\rangle := \sum_{k\in\mathbb{N}} u_k v_k^*$. Let μ be a positive Borel measure on $\mathbb{R}$ satisfying $\int_{\mathbb{R}} x^n d\mu(x) < \infty$ for any $n\in\mathbb{N}$ and $\int_{\mathbb{R}} d\mu(x) = 1$. Let $L^2(\mu)$ denotes the Hilbert space[1] of squared integrable functions on $\mathbb{R}$ with product $\langle a,b\rangle := \int_{\mathbb{R}} a(x)b(x)^* d\mu(x)$.

Definition 2.1 A Jacobi operator J acting on the Hilbert space $\ell^2(\mathbb{N})$ is defined by

$$\begin{aligned}(Je)_m &= a_m e_{m+1} + b_m e_m + a_{m-1}e_{m-1},\ m\geq 1;\\ (Je)_0 &= a_0 e_1 + b_0 e_0,\end{aligned} \tag{2.1}$$

where $\{e_m\}_{m\in\mathbb{N}}$ denotes the canonical orthonormal basis of $\ell^2(\mathbb{N})$ and $a_m\in\mathbb{R}^+$, $b_m\in\mathbb{R}$, $\forall m\in\mathbb{N}$. ♦

By using the explicit expression for the elements of the canonical basis, it is easy to represent (2.1) as an infinite real symmetric tridiagonal matrix with b_m (resp. a_m) $\forall m\in\mathbb{N}$ as diagonal (resp. upper and lower subdiagonals) elements.

The properties of the Jacobi operator depend on the behavior of the sequences $\{a_m\}$, $\{b_m\}$, $m\in\mathbb{N}$. We start by listing simple useful properties:

Proposition 2.2 *Let $\mathcal{D}(\ell^2(\mathbb{N}))$ be the set of finite linear combination of the elements of the canonical basis $\{e_m\}_{m\in\mathbb{N}}$ of $\ell^2(\mathbb{N})$.*
i) J as defined in (2.1) is a symmetric operator with dense domain $\mathcal{D}(\ell^2(\mathbb{N}))$.
ii) If J is bounded, then it extends to a self-adjoint operator on $\ell^2(\mathbb{N})$.

Proof By a simple calculation, one checks that $\langle Ju,v\rangle = \langle u,Jv\rangle$ for any $u,v\in\mathcal{D}(\ell^2(\mathbb{N}))$. Next, one observes that $\mathcal{D}(\ell^2(\mathbb{N}))$ is dense in $\ell^2(\mathbb{N})$. Hence, J is a symmetric operator of $\ell^2(\mathbb{N})$ with dense domain $\mathcal{D}(\ell^2(\mathbb{N}))$ and i) is proven. Let $J^\dagger$ the adjoint of J. When J is bounded, the mere use of standard definitions yields $Dom(J) = Dom(J^\dagger) = \ell^2(\mathbb{N})$ while J and $J^\dagger$ have the same action. Hence $J = J^\dagger$ which proves ii). ■

In the rest of this paper, we will consider only bounded Jacobi operators. Notice that a bounded Jacobi operator is automatically self-adjoint. We recall that not all symmetric densely defined operators have self-adjoint extensions. Their existence is controlled by the deficiency indices (n_+,n_-)[2]. A necessary and sufficient condition for a densely symmetric operator to have self-adjoint extensions is that $n_+ = n_-$ which will be automatically verified by our Jacobi operators

[1] It is understood that 2 functions a and b are identified whenever $\int_{\mathbb{R}} |a(x)-b(x)|^2 d\mu(x) = 0$.

[2] Let T be a densely symmetric operator. The deficiency indices are defined as $n_\pm = \dim\mathcal{N}_{\pm i}$ where $\mathcal{N}_z := \{u\in Dom(T^\dagger) : T^\dagger u = zu\}$ for $z\in\mathbb{C}\backslash\mathbb{R}$

in this paper since they are assumed to be bounded, hence self-adjoint so that their spectrum is real, implying $n_+ = n_- = 0$.

For bounded Jacobi operators, there is a constraint on the related sequences $\{a_m\}$, $\{b_m\}$, $m \in \mathbb{N}$.

Proposition 2.3 *Let J a Jacobi operator defined by the real sequences $\{a_m\}$, $\{b_m\}$, $m \in \mathbb{N}$ as in (2.1). The following properties hold true:*
i) If J is bounded, then, $\{a_m\}_{m\in\mathbb{N}}$ and $\{b_m\}_{m\in\mathbb{N}}$ are bounded sequences.
ii) If $\{a_m\}_{m\in\mathbb{N}}$ and $\{b_m\}_{m\in\mathbb{N}}$ satisfy $\sup_{m\in\mathbb{N}}(|a_m| + |b_m|) \leq M$, then $||J|| \leq 2M$.

PROOF Assume J is bounded. Then, one has for any $u, v \in \mathcal{H}$ $|\langle Ju, v\rangle| \leq ||J||$. This is true for $u = e_k$, $v = e_{k+1}$ and $u = e_k$, $v = e_k$ which using Definition 2.1 yields $|a_m| \leq ||J||$, $b_m \leq ||J||$ which proves i).
Assume now $\sup_{m\in\mathbb{N}}(|a_m| + |b_m|) \leq M$. For any unit vector $u \in \ell^2(\mathbb{N})$, $u = \sum_m u_m e_m$, a simple calculation yields

$$\begin{aligned}
||Ju||^2 &= \sum_m a_{m-1}^2|u_{m+1}|^2 + a_m^2|u_{m-1}|^2 + b_m^2|u_m|^2 + 2a_m b_m(\Re(u_{m-1}u_m^*) + \Re(u_{m+1}u_m^*)) \\
&+ 2a_m a_{m-1}\Re(u_{m-1}u_{m+1}^*) \leq \sum_m a^2(|u_{m+1}|^2 + |u_{m-1}|^2) + b^2|u_m|^2 \\
&+ 2ab(|\Re(u_{m-1}u_m^*)| + |\Re(u_{m+1}u_m^*)|) + 2a^2|\Re(u_{m-1}u_{m+1}^*)| \\
&\leq 2a^2 + b^2 + 2ab(|\Re(u_{m-1}u_m^*)| + |\Re(u_{m+1}u_m^*)|) + 2a^2|\Re(u_{m-1}u_{m+1}^*)| \\
&\leq 2a^2 + b^2 + 4ab|\langle Su, u\rangle| + 2a^2|\langle S^2u, u\rangle| \leq 4a^2 + b^2 + 4ab \leq 4M^2 \qquad (2.2)
\end{aligned}$$

where we set $\sup_m |a_m| = a$, $\sup_m |b_m| = b$ and S is the bounded shift operator defined by $S : \ell^2(\mathbb{N}) \to \ell^2(\mathbb{N})$, $S : e_m \mapsto e_{m+1}$, $\forall m \in \mathbb{N}$ It satisfies obviously $||S|| = 1$ and $|\langle Su, u\rangle| \leq 1$, $|\langle S^2u, u\rangle| \leq 1$. Hence, $||J|| \leq 2M$ and ii) is proven. ■

Now, apply the spectral theorem for bounded operator Theorem A.1 to J. This latter guarantees the existence of a natural (and unique) compactly supported measure rigidely linked to J. Indeed, from Theorem A.1, there is a unique resolution E of the identity on $\ell^2(\mathbb{N})$ such that for any $u, v \in \ell^2(\mathbb{N})$,

$$\langle Ju, v\rangle = \int_{\mathbb{R}} t dE_{u,v}(t). \qquad (2.3)$$

Let U be a Borel set. From Definition A.2, one can define a Borel measure by $U \mapsto E_{u,v} = \langle E(U)u, v\rangle$, for any $u, v \in \ell^2(\mathbb{N})$. Here, e_0 in the canonical basis of $\ell^2(\mathbb{N})$ is a convenient vector to be used by noticing in particular that the set $\{J^n e_0,\ n \in \mathbb{N}\}$ is dense in $\ell^2(\mathbb{N})$, as a standard computation shows, i.e e_0 is cyclic for the action of J. We have an immediate property summarized in the following proposition.

Proposition 2.4 *The measure μ defined by $\mu(U) := E_{e_0,e_0} = \langle E(U)e_0, e_0\rangle$, for any Borel set U is Borel positive.*

PROOF Since $E(U)$ is an orthogonal projection, one has

$$\mu(U) = \langle E(U)^2 e_0, e_0\rangle = \langle E(U)e_0, E(U)e_0\rangle \geq 0. \qquad (2.4)$$

Hence $\mu(U)$ is positive Borel with support $supp(\mu) \subseteq [-||J||, ||J||]$. ■

The fact that μ can be related to polynomials shows up in the following proposition. It stems from the fact one can generate all the vectors of the canonical basis by a suitable polynomial action of J on e_0.

Proposition 2.5 *Let J a bounded Jacobi operator with associated resolution of identity E and $\{e_n\}_{n\in\mathbb{N}}$ the canonical basis of $\ell^2(\mathbb{N})$. For any $n \in \mathbb{N}$, one can find p_n a polynomial of degree n in $\mathbb{R}[X]$ such that*

$$e_n = p_n(J)e_0. \tag{2.5}$$

The spectral measure is completely determined by μ, i.e

$$\langle E(U)e_n, e_m\rangle = \int_U p_n(t)p_m(t)d\mu(t). \tag{2.6}$$

Proof First, one checks by induction that $e_n = p_n(J)e_0$, $\forall n \in \mathbb{N}$ for some polynomials $p_n \in \mathbb{R}[X]$, thanks to $a_n > 0$.
Then, $\langle E(U)e_n, e_m\rangle = \langle E(U)p_n(J)e_0, p_m(J)e_0\rangle = \langle p_m(J)E(U)p_n(J)e_0, e_0\rangle$. By Theorem A.1, J commutes with $E(U)$ and one can write

$$\langle E(U)e_n, e_m\rangle = \langle p_m(J)p_n(J)E(U)e_0, e_0\rangle = \int_U p_n(t)p_m(t)d\mu(t). \tag{2.7}$$

Hence the result. ■

In physics oriented words, the above proposition states simply that the "matrix elements" of the spectral measure are given by the integrals of products of polynomials with respect to some positive compactly supported Borel measure.
From the above Proposition, it should be noted at this level that the spectral measure is entirely determined by μ. In fact, the bridge between families of orthogonal polynomials and bounded Jacobi operators is provided by a corollary of the spectral theorem, known as the Favard theorem whose version for bounded operators will be given in a while. The following standard proposition set up the framework for the relevant orthogonal polynomials and associated 3-term recurrence relation.

Proposition 2.6 *Let $\{P_n\}_{n\in\mathbb{N}}$ be a family of polynomials of $\mathbb{R}[X]$, orthonormal in $L^2(\mu)$ with respect to a compactly supported measure μ. There exists bounded sequences $\{a_n\}_{n\in\mathbb{N}} \subset \mathbb{R}^+$, $\{b_n\}_{n\in\mathbb{N}} \subset \mathbb{R}$ satisfying*

$$\begin{aligned} tP_n(t) &= a_nP_{n+1}(t) + b_nP_n(t) + a_{n-1}P_{n-1}(t),\ k \geq 1 \\ tP_0(t) &= a_0P_1(t) + b_0P_0(t). \end{aligned} \tag{2.8}$$

Proof The fact that $\{P_n\}_{n\in\mathbb{N}}$ is an orthonormal family yields immediately $tP_n(t) = \sum_{p=0}^{n+1} \alpha_pP_p(t)$ with the only non-vanishing α_p's given by $\alpha_{n\pm1} = \int_{\mathbb{R}} tP_{n\pm1}(t)P_n(t)d\mu(t)$, $\alpha_n = \int_{\mathbb{R}} tP_n^2(t)d\mu(t)$. From this follows the recurrence (2.8) with $a_n = \alpha_{n+1}$, $b_n = \alpha_n$, $a_{n-1} = \alpha_{n-1}$ while the positivity of a_n stems from the leading coefficient.
Next, one has

$$|b_n| \leq \int_{\mathbb{R}} |tP_n^2(t)|d\mu(t) \leq (\sup_{t\in supp(\mu)} |t|) \times ||P_n||_{L^2}^2. \tag{2.9}$$

On the other hand

$$|a_n| \leq (\sup_{t\in supp(\mu)} |t|) \int_{\mathbb{R}} |P_{n+1}(t)||P_n(t)|d\mu(t) \leq (\sup_{t\in supp(\mu)} |t|) \times ||P_{n+1}||_{L^2}||P_n||_{L^2}, \tag{2.10}$$

where the last inequality stems from the use of Hölder inequality. From (2.9) and (2.10), one concludes that $\{a_n\}_{n\in\mathbb{N}}$ and $\{b_n\}_{n\in\mathbb{N}}$ are bounded. ■

Remark 2.7 i) Notice that the boundedness of the sequences $\{a_n\}$ and $\{b_n\}$ explicitely uses the fact that the support of μ, *supp*(μ) is compact. This corresponds to some specific families of orthogonal polynomials. This will have to be reconsidered in the case of unbounded Jacobi operators, therefore corresponding to other families of orthogonal polynomials.
ii) Notice also the similarity between (2.8) and (2.1) which signals a link between those Jacobi operators and families of orthogonal polynomials determined by 2 sequences of real numbers as given above. This observation will be strengthened by the Favard theorem for bounded Jacobi operators, given below. ♦

Theorem 2.8 (Favard Theorem) *To each bounded Jacobi operator J is associated a unique compactly supported measure μ satisfying:*
i) Orthonormal set: For any polynomials $P \in \mathbb{R}[X]$, the map $\mathcal{U} : \ell^2(\mathbb{N}) \to L^2(\mu)$ defined by $\mathcal{U} : P(J)e_0 \mapsto P$ extends to a unitary operator and $\{P_n := \mathcal{U}e_n\}_{n\in\mathbb{N}}$ is an orthonormal family with respect to μ.
ii) Intertwiner: Let $L : L^2(\mu) \to L^2(\mu)$ be the left multiplication operator by $t \in \mathbb{R}$. One has $\mathcal{U}J = L\mathcal{U}$.

PROOF First, one has for any $P_1, P_2 \in \mathbb{R}[X]$

$$\langle P_1(J)e_0, P_2(J)e_0\rangle = \langle P_2^\dagger(J)P_1(J)e_0, e_0\rangle = \int_{\mathbb{R}} P_1(t)P_2(t)d\mu(t), \tag{2.11}$$

where the last equality stems from Proposition 2.5 and we use $E(\mathbb{R}) = \mathbb{I}$. Now

$$\int_{\mathbb{R}} P_1(t)P_2(t)d\mu(t) = \langle P_1 P_2\rangle_{L^2} = \langle \mathcal{U}P_1e_0, \mathcal{U}P_2e_0\rangle_{L^2}, \tag{2.12}$$

where the definition of $\mathcal{U}$ has been used to obtain the last equality. Hence $\mathcal{U}^\dagger\mathcal{U} = \mathcal{U}\mathcal{U}^\dagger = \mathbb{I}$. Next, from the beginning of the proof of Proposition 2.5, $e_n = P_n(J)e_0$ implies that the subspace $\{J^n e_0,\ n \in \mathbb{N}\}$ of $\ell^2(\mathbb{N})$ on which $\mathcal{U}$ acts is dense in $\ell^2(\mathbb{N})$ while the image of $\mathcal{U}$ is dense in $\mathbb{R}[X]$ because *supp*(μ) is compact. Hence $\mathcal{U}$ extends to a unitary operator $\mathcal{U} : \ell^2(\mathbb{N}) \to L^2(\mu)$. Finally, simply write

$$\int_{\mathbb{R}} P_n(t)P_m(t)d\mu(t) = \langle \mathcal{U}e_n, \mathcal{U}e_m\rangle_{L^2} = \langle e_n, e_m\rangle = \delta_{nm}, \tag{2.13}$$

where we used the fact that $\mathcal{U}$ is unitary and i) is proven.
Now, from a standard computation using $\mathcal{U}Je_n = L\mathcal{U}e_n$ for any $n \in \mathbb{N}$ together with Propositions 2.5, 2.6 and Definition 2.1, one obtains easily

$$a_n = \langle Je_n, e_{n+1}\rangle = \int_{\mathbb{R}} tP_n(t)P_{n+1}(t)d\mu(t),\ b_n = \langle Je_n, e_n\rangle = \int_{\mathbb{R}} tP_n^2(t)d\mu(t). \tag{2.14}$$

Hence, ii) holds true.
Finally, let $s_n := \int_{\mathbb{R}} x^n d\mu(x)$, $n \in \mathbb{N}$, be the moments of μ with the corresponding Stieltjes transform $w(z) = \int_{\mathbb{R}} \frac{1}{x-z}d\mu(x)$, for any $z \in \mathbb{C}\backslash\mathbb{R}$. Using the fact that μ is compactly supported, $supp(\mu) = [-\sigma, \sigma]$, one can write $w(z) = -\frac{1}{z}\sum_{n=0}^{\infty}\int_{\mathbb{R}}(\frac{x}{z})^n = -\sum_{n=0}^{\infty}\frac{s_n}{z^{n+1}}$ where the RHS is absolutely convergent whenever $|z| > \sigma$. Then, further using the formula

$$\lim_{\delta\to 0} \pi^{-1}\int_a^b \Im[w(t + i\delta)]dt = \mu([a, b]) + \frac{1}{2}(\mu(a) + \mu(b)), \tag{2.15}$$

one verifies that μ is determined uniquely by its moments. The theorem is proven. ■

Remark 2.9 i) From Proposition 2.4 and 2.5, one infers $\langle E(U)e_0, e_0\rangle = \int_U P_0(t)P_0(t)d\mu(t)$. By choosing $U = \mathbb{R}$, one obtains $\int_{\mathbb{R}} d\mu(t) = 1$ provided $P_0 = 1$. Therefore, this last normalization condition, combined to the 3-term recurrence (2.8) of Proposition 2.6 completely determines the set of polynomials P_n and ensures that μ is a probability measure.
ii) The Favard theorem extends to the case of unbounded Jacobi operators, up to moderate modifications.
iii) We point out once more time that the above measure μ is entirely determined from the full sequence of its moments, $\{s_n := \int_{\mathbb{R}} x^n d\mu(x)\}_{n\in\mathbb{N}}$. This deals with the so-called moment problem, another important block involved in the theory of orthogonal polynomials, which however is not directly used in the present paper. For more details, see e.g in [8], [20]. ♦

In the ensuing analysis, we will use at some places the Christoffel-Darboux relations. There are other algebraic formulas but we will not need them in this paper.

Proposition 2.10 *Let f' denotes the derivative with respect to x. The following relation holds true:*

$$(t-z)\sum_{k=0}^{n} P_k(t)P_k(z) = a_n(P_{n+1}(t)P_n(z) - P_n(t)P_{n+1}(z)), \tag{2.16}$$

$$\sum_{k=0}^{N-1}(P_k(x))^2 = P'_N(x)P_{N-1}(x) - P'_{N-1}(x)P_N(x). \tag{2.17}$$

PROOF This is routine computation. Multiply (2.8) by $P_n(z)$, then (2.8) in the z variable by $P_n(t)$ and take the sum. ■

Let us use a more physics oriented language for a while: In the following, the "diagonalization" of the self-adjoint (positive) kinetic operator of the considered NCFT will give rise to a 3-term recurrence equation satisfied by the "matrix elements" of a unitary (orthogonal) operator. This will therefore be related to a given family of orthogonal polynomials.

Let us make a direct connection with NCFT for which the computation of the propagator may be a terrible task when carried out by brute force. But this computation can be actually simplified by using the following simple technical lemma which is a direct consequence of the Favard theorem together with the continuous functional calculus. It gives rise easily to explicit expression for the "matrix elements" of the propagator, once the family of orthogonal polynomials has been identified. Notice that it can be easily generalized to other situations (e.g $\ell^n(\mathbb{N})$ and/or unbounded self-adjoint Jacobi operators). Assume that the kinetic operator for a NCFT J is a (bounded) Jacobi operator as in Definition 2.1 with J^{-1} such that $\{0\} \notin spec(J)$. According to the material we have presented above in this section, the following relations hold true for any $m, n \in \mathbb{N}$:

$$\langle e_m, Je_n\rangle = \int_{supp(\mu)} tP_m(t)P_n(t)d\mu(t), \tag{2.18}$$

$$\langle e_m, J^{-1}e_n\rangle = \int_{supp(\mu)} \frac{1}{t}P_m(t)P_n(t)d\mu(t). \tag{2.19}$$

where of course $\{P_n\}_{n\in\mathbb{N}}$ is the unique family of orthogonal polynomials associated to J. These relations are direct consequences of Definition 2.1, Proposition 2.6 and e.g Proposition 2.5.

To close this subsection, it is usefull to point out the relationship between the spectrum of a truncated J, i.e restricted to a $N \times N$ matrix, and the zeros of the related orthogonal polynomials. This will be helpfull in the process of characterization of the total spectrum of the kinetic operators, i.e point (discrete), continuous and residual spectra.

Lemma 2.11 *For a given $N \in \mathbb{N}$ let J^N denotes the matrix of $\mathbb{M}_N(\mathbb{R})$ obtained from J (2.1) by restricting the indices $m, n, ...$ to $0 \le m, n, ... \le N-1$. Then, the eigenvalues of J^N are exactly given by the zeroes of the orthogonal polynomial P_N in the orthogonal family defined by (2.8) related to J. The zeros of the orthogonal polynomials are simple and real.*

Proof For $m \le N$, the 3-term recurrence (2.8) can be easily cast into the matrix form

$$t\mathcal{P}_N(t) = J^N \mathcal{P}_N(t) + a_N P_N(t) e_N,\ \ \mathcal{P}_N := (P_0(t), P_1(t), ..., P_{N-1}(t)). \tag{2.20}$$

Whenever $t = t_0$, a zero of $P_N(t)$, (2.20) becomes $t_0\mathcal{P}_N(t_0) = J^N\mathcal{P}_N(t_0)$. Now, J^N is symmetric tridiagonal real matrix so it has real, simple eigenvalues. The lemma is proven. ■

Many (but not all) properties presented here for the case of bounded Jacobi operators will still apply (up to minor adaptations) to the case of unbounded operators. Note by the way that the zeros of the orthogonal polynomials for bounded Jacobi all belong to $[-||J||, ||J||]$ as a corollary of Lemma 2.

The case of NCFT with unbounded Jacobi operators is slightly more complicated and will be presented elsewhere [22]. One difference comes from the "large momentum" behavior of the inverse of the kinetic operator, i.e the propagator, which decays in the latter case while it may remain non decaying in the bounded case, implying the occurrence of non zero correlation at large separation of indices in the language of matrix bases.

However (at least in the classes of noncommutative spaces we considered) the differences have a milder effect on the associated spectral triples (i.e basically the spectral triple with the square root of the kinetic operator as Dirac operator from which the NCFT action would derive as a spectral action). This comes mainly from the fact that the relevant non unital algebras are represented on a smooth (e.g Hilbert-Schmidt) subalgebra of some algebra of bounded operators. The latter representation enters explicitely the definition of some localized axioms for the non unital triple: functionals of the Dirac operator are (left or right) multiplied by smooth (e.g Hilbert-Schmidt) operator. This facilitates the obtention of compactness or summability conditions. In this respect, the bounded Jacobi operator case is not "too far away" from the unbounded case and is representative of common features. In this paper, the kinetic operator of the NCFT we consider is bounded Jacobi.

3. Gauge theories on $\mathbb{R}^2_\theta$ as matrix models

3.1. Noncommutative gauge theories: A review on a nutshell

In this subsection, we will focus on some of the open problems so far left unsolved within noncommutative gauge theories built on what could be called informally "totally noncommutative geometries". We will not consider the recent fascinating developments in gauge models on "almost commutative geometries, in particular gauge models of Connes-Chamseddine-types that reproduce the main quantitative features of the Standard Model. See e.g 2nd of ref. [1], [34], [35]and references therein.

The NCFT studied in this paper is rigidely linked to a gauge theory on the noncommutative Moyal plane. It is therefore instructive to summarize the present situation for noncommutative gauge theories.

At the classical level, the construction of gauge invariant actions is not so difficult, once a differential calculus has been set up, together with a proper notion of noncommutative connection. This basically amounts to built a gauge invariant polynomial depending on some curvatures, usually a trace of products of curvatures. The use of spectral triples provides a natural way to construct noncommutative differential calculi as well as natural gauge invariant actions, once the spectral action principle is accepted. Another way is to use directly specific versions of the differential calculus based on derivations of the associative algebra related to the noncommutative space. This more algebraic way is often used in the mathematical physics literature as a flexible and fast tool to obtain expressions for noncommutative actions of NCFT. It extends easily to the construction of gauge invariant actions defined on "noncommutative spaces" and it appears that most of the developments in noncommutative gauge theories fit well within definite versions of derivation based differential calculi with the notion of noncommutative connection being a natural extension of the Koszul connection. This noncommutative differential calculus has been introduced a long ago in [36] (see also [37] and references therein). For further more recent developments and extensions see [38], [39], [40].

Perturbative renormalisation of noncommutative gauge theories involves so far unsolved problems. Renormalisation of NCFT is in general more difficult than for commutative (local) field theories. This is often due to the UV/IR mixing. It occurs in the simplest noncommutative φ^4 model on the 4-d Moyal space, as noticed in [41], and destroys renormalisability. The phenomenon persists in noncommutative gauge models and is one of the main unsolved problems within Moyal NCFT. A 1st solution to this problem leading to renormalisable NCFT has been proposed in 2003 [9], [10]. It is obtained by supplementing the scalar action with a harmonic oscillator term, leading to the now popular Grosse-Wulkenhaar model, whose 4-d version is likely non-perturbatively solvable, see [42]. The Grosse-Wulkenhaar model together with its gauge counterpart, to which we will turn on below, is actually related to an interesting NCG, called finite volume spectral triple and analyzed in [27]. It turns out the related noncommutative metric geometry is homothetic in a very precise sense to the usual noncommutative Moyal metric geometry, as shown in [28].

At the present time, no (all orders) renormalisable gauge theory on 4-d Moyal space has been constructed, despite various encouraging progresses in that direction. Inspired by the above Grosse-Wulkenhaar solution, a gauge model obtained either by heatkernel methods or by (equivalent) effective action computation, starting from of the Grosse-Wulkenhaar model coupled to a gauge field, gave rise to what is sometimes called the "induced gauge theory" on Moyal space [23], [24]. The corresponding action is (the notations are by now standard):

$$S = \int d^4x \Big(\frac{1}{4} F_{\mu\nu} \star F_{\mu\nu} + \frac{\Omega^2}{4} \{\mathcal{A}_\mu, \mathcal{A}_\nu\}^2_\star + \kappa \mathcal{A}_\mu \star \mathcal{A}_\mu \Big) \tag{3.1}$$

in which the 2nd and 3rd terms may be viewed as "gauge counterparts" of the harmonic term of the Grosse-Wulkenhaar model, while the first term that looks like a Yang-Mills action has bad UV/IR mixing already showing up as a hard IR singularity in the vacuum polarization tensor. Here, $\mathcal{A}_\mu$ is the covariant coordinates, a natural gauge covariant tensor form stemming from the existence of a canonical gauge invariant connection in the present NC framework. We refer to [38], [39] for a complete description of the underlying noncommutative differential calculus and relevant notion of noncommutative connection. Here, it is sufficient to note that the relevant gauge transformation are

$$\mathcal{A}_\mu \to \mathcal{A}_\mu^g = g^\dagger \star \mathcal{A}_\mu \star g, \ \forall g \in \mathcal{U}(\mathbb{R}^4_\theta) \tag{3.2}$$

with similar covariant transformation for the curvature $F_{\mu\nu}$. Here $\mathbb{R}^4_\theta$ is identified in this algebraic framework to the multiplier algebra of the Schwartz functions algebra with associative Moyal product and $\mathcal{U}(\mathbb{R}^2_\theta)$ denotes the corresponding unitary elements. The structure of the (symmetric) vacua of (3.1) has been analyzed in detail and classified in [25]. It was realized from this work that the complicated vacuum structure thus exhibited forbids a direct perturbative treatment of the gauge action (3.1).

Alternative way of introducing a Grosse-Wulkenhaar term within a gauge invariant theory has been also proposed. These are based on the implementation of a IR damping mechanism [43]. In particular in the 1st ref. of [43], it has been claimed that renormalisability of a noncommutative gauge-invariant model could be restored by building an IR damping into the propagator for the gauge potential propagator together with nonlocal counterterms related to the (quadratic and linear) IR singularities triggering the UV/IR mixing, akin to the "soft-breaking" terms of the Gribov-Zwanziger action. In [44], Slavnov-Taylor identities have been used to track the IR singularities in gauge models of the above type, showing basically that the needed independent Gribov-type parameters reduce in fact to one. This damping approach is appealing. However, interpreting the action within the framework of a definite noncommutative differential geometry is unclear at the present time, unlike the case of the induced gauge action.

We mention finally that modifications of the noncommutative differential calculus underlying the action (3.1) have been investigated in order to tackle the UV/IR mixing, resulting in modifications of (3.1) [38, 39]. Although interesting relationships have been shown between these modifications and the Grosse-Wulkenhaar model, no improvement of the UV/IR mixing behavior has been obtained. Notice that the inclusion of fermions is expected to improve the situation.

Another appealing (in some sense dual) approach is provided by the matrix model formulation of noncommutative gauge theory has also been investigated, initiated a long ago in [45] in the context of type IIB (stringy) matrix models. This basically amounts to re-interpret the noncommutative gauge theories as matrix models taking advantage of the relationship between the gauge potential and the covariant coordinate mentionned above, exhibiting in some cases a relationship with the action (3.1). For recent exhaustive reviews, see [46] (see also [47]). Note that the matrix model approach may in some cases allow one to go beyond the perturbative approach [48]. One interesting outcome is that it may provide a interpretation for the UV/IR mixing in some noncommutative gauge theories in terms of an induced gravity action. See e.g [49].

Motivated by this alternative approach, the induced gauge theory action (3.1) on the Moyal plane, expanded around a particular symmetric vacuum, has been formulated as a matrix model and its 1-loop behavior has been studied [26]. This analysis provides the first investigation of the quantum properties of (3.1), showing by the way the usefulness of the matrix model formulation to overcome the long standing difficulty linked to the vacuum structure. One of the result is the appearance of a non vanishing 1-point (tadpole) function indicating a quantum vacuum instability. Regardless its gauge theory origin, this model is essentially a NCFT with a particular bounded Jacobi operator as kinetic operator and polynomial interactions after suitable gauge fixing and is actually representative of most of the main features of any other NCFT with other bounded Jacobi kinetic operator. Note also that some (but not all!) features of this bounded case will be somewhat similar to the unbounded case. This is why we have chosen to investigate deeply the above quantum matrix gauge theory on the Moyal plane. This will be done in the sequel. Notice by the way that this gauge model is related to the 6-vertex model (to which it reduces whenever the vacuum is the trivial one [26]).

More recently, we have investigated gauge theories on $\mathbb{R}^3_\lambda$ [14]. While non-vanishing 1-point function also occurs at the one-loop order, deserving further investigation, it appears that the UV behavior as well as the IR one are milder than what one would obtain for a NCFT on (4-d) Moyal space. This behavior was also observed in the families of scalar NCFT on $\mathbb{R}^3_\lambda$ considered in [13], among which classes of finite NCFT have been identified, leading to the conclusion that no perturbatively dangerous UV/IR mixing should occur within these scalar NCFT. One could object that it just comes from dimensional effect as it happens for commutative field theories and/or that it reflects a particular choice of the differential calculus. This is not correct. In fact, a closer inspection of inspection of $\mathbb{R}^3_\lambda$ shows that it has a natural decomposition as $\mathbb{R}^3_\lambda \simeq \oplus_{j\in\frac{\mathbb{N}}{2}}\mathbb{M}_{2j+1}(\mathbb{C})$, reflecting the fact that $\mathbb{R}^3_\lambda$ is a closure of $U(su(2))$, the universal envelopping algebra of $su(2)$. This yields one-loop amplitudes in which j, the radius of the fuzzy sphere $\mathbb{S}_j = \mathbb{M}_{2j+1}(\mathbb{C})$ in the physicists language, plays the role of a natural UV

Journal of Physics: Conference Series **634** (2015) 012006 doi:10.1088/1742-6596/634/1/012006

cut-off. Two important comments are in order. First, the models considered in [13] and [14] are *not* models on fuzzy sphere. Only suitable truncations (reduction in the physicists language) of these models to a single fuzzy sphere are! Next, the noncommutative differential calculus used in [13] and [14], although perfectly admissible, is not the most natural one. In fact, there is a natural action of $\mathbb{C}(SU(2))$ onto $\mathbb{R}^3_\lambda$, stemming from the duality pairing between $U(su(2))$ and $\mathbb{C}(SU(2))$. This yields a bicovariant differential calculus on $\mathbb{R}^3_\lambda$ with a Laplacian whose commutative limit is the usual Laplacian on $\mathbb{R}^3$. These observations identify ways of future investigations. It should be interesting to see in what extend the above mild UV and IR behavior is affected by the change of differential calculus and to see how this behavior survives at higher orders in perturbation. It would then favor an interesting more general class of noncommutative spaces leading to nicely behaving NCFT, to which $\mathbb{R}^3_\lambda$ belongs as an element. Notice that the gauge theory on $\mathbb{R}^3_\lambda$ studied in [14] when reduced to a single fuzzy sphere in $\mathbb{R}^3_\lambda$, can be identified with the Alekseeev-Recknagel-Schomerus action [15] describing some low energy action for brane dynamics on $\mathbb{S}^3$.

3.2. Quantum matrix gauge theory on $\mathbb{R}^2_\theta$

We begin this subsection by synthesizing the properties of the Moyal plane that are actually needed in the subsequent analysis. There is a huge literature on Moyal spaces. For a complete presentation, see in e.g [2] and references therein.

Informally, the Moyal plane can be viewed as a strict quantization of $\mathbb{R}^2$ to the Poisson bracket on $C_0(\mathbb{R}^2)$. Behind the common vocable "Moyal algebra", there are various algebras depending on the required properties of regularity and/or differentiability. To deal with NCFT, it is especially convenient to view the relevant Moyal algebra as already represented on a suitable C^*-algebra of bounded operators on some Hilbert space. This is the viewpoint we adopt here, which permits one to avoid cumbersome manipulations of integrals of star-products arising when working with (Frechet) algebras of functions. The simplification is especially efficient in the case of the noncommutative space $\mathbb{R}^3_\lambda$ for which the measure in the relevant integral is not the usual Lebesgues measure on $\mathbb{R}^3_\lambda$, but has a complicated expression.

Let $\mathcal{S}$ be the space of $\mathbb{C}$-valued Schwartz functions on $\mathbb{R}^2$ and $\{f_{mn}(x)\}_{m,n\in\mathbb{N}} \subset \mathcal{S} \subset L^2(\mathbb{R}^2)$ be the family of Wigner transition eigenfunctions of the 1-d harmonic oscillator, with $(\theta > 0)$

$$f_{mn} \star f_{kl} = \delta_{nk} f_{ml},\ f^\dagger_{mn} = f_{nm},\ \langle f_{mn}, f_{kl}\rangle_{L^2} = 2\pi\theta\delta_{mk}\delta_{nl}, \tag{3.3}$$

where $\star$ is the Moyal product whose explicit expression will not be needed here. Let $(e_n)_{n\in\mathbb{N}}$ be the canonical basis of $\ell^2(\mathbb{N})$, see subsection 2. First, define the following Frechet $*$-algebras

$$\mathbb{A} : = \{\phi \in (\mathcal{S}, \star),\ \rho_{\alpha,\beta}(\phi) = \sup_{x\in\mathbb{R}^2} |x_1^{\alpha_1} x_2^{\alpha_2} \partial_1^{\beta_1} \partial_2^{\beta_2} \phi| < \infty,\ \forall \alpha_i, \beta_i \in \mathbb{R}^+\} \tag{3.4}$$

$$\mathcal{M}_\theta : = \{(\Phi_{m,n})_{m,n\in\mathbb{N}},\ \rho_n(\Phi) = \sum_{p,k} \theta(p+\frac{1}{2})^n (k+\frac{1}{2})^n |\Phi_{pk}|^2 < \infty,\ \forall n \in \mathbb{N}\}, \tag{3.5}$$

i.e (3.5) is the Frechet subalgebra of $\ell^2(\mathbb{N}^2)$ involving the rapid decay matrices (equipped with matrix product).

It is known that $\mathbb{A}$ is isomorphic to $\mathcal{M}_\theta$ as Frechet $*$-algebras, with isomorphism defined by $\Phi_{mn} \mapsto \sum_{m,n} \Phi_{mn} f_{mn} \in \mathcal{S}$ and inverse $\phi \in \mathcal{S} \mapsto \frac{1}{2\pi\theta}\langle \Phi, f_{mn}\rangle_{L^2}$. Next, $\mathcal{M}_\theta$ is represented on a subalgebra of the C^*-algebra $\mathcal{B}(\ell^2(\mathbb{N}))$ by using the natural representation of matrices as operators on $\ell^2(\mathbb{N})$, denoted by η. This is simply the product of a matrix by a column vector. Indeed, for any $\Phi \in \mathcal{M}_\theta$, $\Phi = \sum_{m,n} \Phi_{mn} e_m \otimes e_n$, define

$$\eta : \ell^2(\mathbb{N}) \otimes \ell^2(\mathbb{N}) \to \mathcal{B}(\ell^2(\mathbb{N})),\ \eta(e_m \otimes e_n) = e_m \otimes e^*_n,\ \forall m,n \in \mathbb{N}$$

with dual e_n^* such that $e_n^*(e_p)=\delta_{np}$. In order to make contact with the physicists notations, we will set $\eta(e_m\otimes e_n)=e_m\otimes e_n^*:=|m><n|=f_{mn}$ in a while.

The definition of $\mathcal{M}_\theta$ (3.5) implies that the operator $\eta(\Phi)$ is Hilbert-Schmidt (set $n=0$ in the seminorm $\rho_n(\Phi)$). Hence $\eta(\Phi)$ is compact. Thus $\eta(\mathcal{M}_\theta)$ inherits a C^*-norm from its Hilbert-Schmidt action on $\ell^2(\mathbb{N})$ (and so does $\mathcal{M}_\theta$). One concludes that the closure of $\eta(\mathcal{M}_\theta)$ is such that $\overline{\eta(\mathcal{M}_\theta)}\subseteq\mathcal{K}(\ell^2(\mathbb{N}))$ where $\mathcal{K}(\ell^2(\mathbb{N}))$ is the C^*-subalgebra of compact operators on $\ell^2(\mathbb{N})$. But since η is faithfull, one has the isomorphism $\eta(\mathcal{M}_\theta)\simeq\mathcal{M}_\theta$. Hence $\overline{\mathcal{M}_\theta}\subset\mathcal{K}(\ell^2(\mathbb{N}))$. On the other hand, $\overline{\mathcal{M}_\theta}\supseteq\mathcal{K}(\ell^2(\mathbb{N}))$ since $\mathcal{M}_\theta\supseteq\bigcup_{n=1}^{\infty}\mathbb{M}_n(\mathbb{C})$ holds true. Thus $\overline{\mathcal{M}_\theta}=\mathcal{K}(\ell^2(\mathbb{N}))$. Finally, the map $U:L^2(\mathbb{R}^2)\to\ell^2(\mathbb{N})\otimes\ell^2(\mathbb{N})$ defined by $U:f_{mn}\mapsto\sqrt{2\pi\theta}e_m\otimes e_n$ for any $m,n\in\mathbb{N}$ is an isometry. Hence $\bar{\mathbb{A}}\simeq\mathcal{K}(\ell^2(\mathbb{N}))$. Thus, we are dealing with compact operators at the end of the day.

In the following, it is understood that the algebra $\mathbb{A}$ is already represented on $\mathcal{B}(\ell^2(\mathbb{N}))$ as we have explained above. The resulting algebra will be sometimes denoted by $\mathbb{R}^2_\theta$ for short. According to the above discussion, we drop the η symbol everywhere and denote any element in $\mathbb{R}^2_\theta$ (that corresponds actually to a field in the NCFT framework) as $\Phi=\sum_{m,n\in\mathbb{N}}\Phi_{mn}|m><n|$, where the $|m><n|$'s correspond to the representation of the f_{mn}'s on $\mathcal{B}(\ell^2(\mathbb{N}))$. To simplify the notations, we will from now on set $|m><n|=f_{mn}$ (i.e use the same notation for, says operators and corresponding symbols). Of course, the Moyal $\star$-product becomes formally a "matrix product" (since η is a $*$-algebra morphism, $\eta(MN)=\eta(M)\eta(N)$) and the above relations for the f_{mn}'s still hold true among the represented f_{mn} while

$$\int_{\mathbb{R}^2}\to\text{tr}\,.$$

Note that the Moyal product can be extended to larger algebras, by exploiting families of seminorms, see e.g [52] for an extensive discussion.

In the following, we use the covariant coordinate $\mathcal{A}_\mu$ defined above as the fundamental field variable in the action, i.e $S=S[\mathcal{A}_\mu]$ which amounts to interpret the induced gauge theory (3.1) as a matrix model. This latter is invariant under the gauge transformations $\mathcal{A}_\mu\to\mathcal{A}_\mu^g=g^\dagger\star\mathcal{A}_\mu\star g$, $\forall g\in\mathcal{U}(\mathbb{R}^2_\theta)$. Notice by the way the formal similarity between the 1st term of the action (3.1) at $\Omega=0$ and the bosonic part of the IKKT matrix model (see e.g [51] and references therein). The suitable differential calculus underlying the present functional action can be viewed as the simplest differential calculus à la Koszul generated by the 2 natural derivations of $\mathbb{R}^2_\theta$. This has been discussed and analyzed at length in [38, 39] to which we refer for details. The only tools we need for the moment are the gauge transformations of the covariant coordinates given above and general features of the natural basis (f_{mn}), $m,n\in\mathbb{N}$ of $\mathbb{R}^2_\theta$, summarized by (3.3). The 2-dimensional version of (3.1) is

$$S_\Omega[\mathcal{A}]=\text{tr}\left((1+\Omega^2)\mathcal{A}\mathcal{A}^\dagger\mathcal{A}\mathcal{A}^\dagger+(3\Omega^2-1)\mathcal{A}\mathcal{A}\mathcal{A}^\dagger\mathcal{A}^\dagger+2\kappa\mathcal{A}\mathcal{A}^\dagger\right),\tag{3.8}$$

where we define $\mathcal{A}=\frac{\mathcal{A}_1+i\mathcal{A}_2}{\sqrt{2}}$, $\mathcal{A}^\dagger=\frac{\mathcal{A}_1-i\mathcal{A}_2}{\sqrt{2}}$. Notice that the functional action (3.8) has some similarities with the 6-vertex model[3]. The corresponding equation of motion is

$$(3\Omega^2-1)(\mathcal{A}^\dagger\mathcal{A}\mathcal{A}+\mathcal{A}\mathcal{A}\mathcal{A}^\dagger)+2(1+\Omega^2)\mathcal{A}\mathcal{A}^\dagger\mathcal{A}+2\kappa\mathcal{A}=0.\tag{3.9}$$

We now have to choose a particular vacuum (the background), expand the action around it, fix the background symmetry of the expanded action. Focusing on vacuum configurations with

[3] This similarity has been pointed out to us by H. Steinacker. However, the actual analysis and the actual "identity" of the model depends on the choice of a particular vacuum around which the classical theory is expanded, as illustrated below.

Journal of Physics: Conference Series **634** (2015) 012006 doi:10.1088/1742-6596/634/1/012006

the global symmetries of the classical action as in [25] amounts to look for solutions Z of the equation of motion with the following expansion

$$Z = \sum_{m,n\in\mathbb{N}} Z_{mn} f_{mn},\ Z_{mn} = -ia_m\delta_{m+1,n}, \forall m,n \in \mathbb{N} \tag{3.10}$$

where the sequence of complex numbers $\{a_m, m \in \mathbb{N}\}$ satisfies

$$a_m\big[(3\Omega^2-1)(|a_{m+1}|^2+|a_{m-1}|^2)+2(1+\Omega^2)|a_m|^2+2\kappa\big]=0, \tag{3.11}$$

obtained by plugging Z (3.10) into (3.9). The trivial vacuum solution is $a_m = 0$. The corresponding model would then be (a version of) the 6-vertex. We turn now on the non trivial symmetric vacua classified in [25]. They are given, up to an unessential phase (ξ_m in the notations of [25]), by $a_m = \sqrt{u_m}$ with

$$0<\Omega^2<\frac{1}{3}\ ,\quad u^1_m=\alpha(r^m-r^{-m})-\frac{\kappa}{4\Omega^2}(1-r^{-m}),\ \alpha\geq 0,\ r>1 \tag{3.12}$$

$$\Omega^2=\frac{1}{3}\ ,\quad \kappa<0,\ a_m=\frac{1}{2}\sqrt{-3\kappa},\ \forall m\in\mathbb{N} \tag{3.13}$$

$$\frac{1}{3}<\Omega^2<1\ ,\quad u^2_m=-\frac{\kappa}{4\Omega^2}(1-r^{-m}),\ \kappa\leq 0,\ r\leq -1 \tag{3.14}$$

$$\Omega^2=1\ ,\quad u^3_m=-\frac{\kappa}{4}(1-(-1)^{-m}),\ \kappa\leq 0 \tag{3.15}$$

where

$$r=\frac{1+\Omega^2+\sqrt{8\Omega^2(1-\Omega^2)}}{1-3\Omega^2}. \tag{3.16}$$

Within the present phase choice, $a_m \in \mathbb{R}$. Let us summarize the main steps of the derivation of the gauge fixed action performed in [26]. By setting formally $\mathcal{A} = Z + \phi$, $\mathcal{A}^\dagger = Z^\dagger + \phi^\dagger$ into (3.8), where ϕ can be interpreted as a fluctuation around Z, one easily obtains a functional action $S[\phi,\phi^\dagger]$ invariant under a background transformation related to a nilpotent BRST-like operation, δ_Z, with structure equations:

$$\delta_Z\phi=-i[Z+\phi,C],\ \delta_Z\phi^\dagger=-i[Z^\dagger+\phi^\dagger,C],\ \delta_Z Z=0,\ \delta_Z C=iCC. \tag{3.17}$$

Here, C, $\bar{C}$ and b are respectively the ghost, the antighost and the Stückelberg field with ghost number equal to $+1$, -1 and 0. δ_Z acts as a graded derivation with grading equal to the sum of the degree of forms and ghost number (modulo 2) and $\delta_Z^2 = 0$. This background symmetry can be fixed by supplementing $S[\phi,\phi^\dagger]$ with the gauge-fixing action

$$S_{GF}=\delta_Z\,\mathrm{tr}\ \bar{C}(\phi-\phi^\dagger)=\mathrm{tr}\ (b(\phi-\phi^\dagger)+i\bar{C}[Z-Z^\dagger+\phi-\phi^\dagger,C]), \tag{3.18}$$

where $\delta_Z\bar{C} = b$, $\delta_Z b = 0$. There are interesting underlying algebraic structures which have been investigated in [50]. Upon integrating the b field, the ghost fields decouple. Setting now $\phi = \sum_{m,n}\phi_{mn}f_{mn}$, the remaining (non-ghost) part of the gauge fixed action reduces to a functional of ϕ only given by

$$S[\phi]=\sum_{m,n,k,l\in\mathbb{N}}\phi_{mn}\phi_{kl}G_{mn;kl}+S_{int} \tag{3.19}$$

where the kinetic operator takes the complicated expression

$$\begin{aligned}G_{mn;kl} &= (1+5\Omega^2)\delta_{ml}\delta_{nk}(a_na_{n+1}+a_na_{n-1})\\ &- (3\Omega^2-1)(\delta_{ml}\delta_{n+1,k-1}a_na_{n+1}+\delta_{ml}\delta_{n-1,k+1}a_na_{n-1}-2\delta_{m,l+1}\delta_{k+1,n}a_na_l)\\ &- (1+\Omega^2)(\delta_{k,n+1}\delta_{m,l+1}a_na_l+\delta_{n,k+1}\delta_{l,m+1}a_na_l)+2\kappa\delta_{ml}\delta_{nk}\end{aligned} \tag{3.20}$$

with the a_m's given by any of the sequences defined in (3.12)-(3.16).

The kinetic operator involves 2 different types of terms: i) terms proportional to $(3\Omega^2 - 1)$ which are non zero when $m + n = l + k \pm 2$, ii) terms that are non zero when $m + n = l + k$ which is similar to the so called conservation law for indices of the Grosse-Wulkenhaar model. The cubic and quartic interaction terms are

$$S_{int} = 8\Omega^2 \sum_{m,p,q,r\in\mathbb{N}} i\phi_{pq}\phi_{qr}\phi_{mp}(a_r\delta_{m+1,r} - a_r\delta_{r+1,m}) + 4\Omega^2 \sum_{m,n,k,r\in\mathbb{N}} \phi_{mn}\phi_{nk}\phi_{kr}\phi_{rm}. \quad (3.21)$$

At this point, one remark is in order. Eqn. (3.20) shows clearly that the kinetic operator, hence the dynamics coded by the model, depends essentially on the choosen vacuum.

From now on, we choose the vacuum given by (3.13), namely:

$$\Omega^2 = \frac{1}{3},\ \kappa < 0,\ a_m = \frac{1}{2}\sqrt{-3\kappa},\ \forall m \in \mathbb{N}.$$

for which the analysis is slightly simpler. The kinetic operator (3.20) becomes

$$G^{(1/3)}_{mn;kl} = (-\kappa)\big(2\delta_{ml}\delta_{nk} - \delta_{k,n+1}\delta_{m,l+1} - \delta_{n,k+1}\delta_{l,m+1}\big), \quad (3.23)$$

and fullfills

$$G^{(1/3)}_{mn;kl} \neq 0 \iff m + n = k + l. \quad (3.24)$$

Then, upon setting $n = \alpha - m$, $k = \alpha - l$, with $\alpha = m + n = k + l$ into (3.23), the indice conservation law (3.24) leads to

$$G^{(1/3)}_{m,\alpha-m;\alpha-l,l} := G^{\alpha}_{m,l} = \mu^2(2\delta_{ml} - \delta_{m,l+1} - \delta_{l,m+1}),\ \forall m, l \in \mathbb{N} \quad (3.25)$$

with $\mu^2 := -\kappa$, which is independent of α. Some (spectral) properties of the kinetic operator defined by (3.25) can now be easily obtained by a mere application of the material developed in subsection 2.

Proposition 3.1 *Let $G_{nm} := G^{\alpha}_{n,m}$ be the kinetic operator on $\ell^2(\mathbb{N})$ defined by (3.25). The following properties hold true:*
i) The operator on $\ell^2(\mathbb{N})$ given by $(-G)_{nm}$ defines a bounded (thus self-adjoint) Jacobi operator on $\ell^2(\mathbb{N})$.
ii) The spectrum of the truncated kinetic operator G^N_{nm}, $0 \le n, m \le N - 1$ is given by

$$spec(G^N) = \{\lambda^N_k := 2\mu^2(1 - \cos(\frac{(k+1)\pi}{N+1})),\ k \in \{0, 1, ..., N-1\}, N \in \mathbb{N}^*\}.. \quad (3.26)$$

iii) The spectrum of the kinetic operator G is simple.

Proof Set $J := -\frac{1}{\mu^2}G$. Using (3.25), one has $(Je)_n = e_{n+1} + e_{n-1} - 2e_n$, $(Je)_0 = e_1 - 2e_0$. Then, simply use Definition 2.1 and Propositions 2.2 and 2.3 with $a_n = 1$, $b_n = -2$ for any $n \in \mathbb{N}$ and i) is proven.
To prove ii), one first determine the associated family of orthogonal polynomials. Then, Favard Theorem 2.8 and Proposition 2.6 guaranty the existence of a family of orthogonal polynomials with respect to a unique compactly supported probability measure with 3-term recurrence given by

$$P_{n+1}(t) + P_{n-1}(t) - 2P_n(t) = tP_n(t),\ P_1(t) - 2P_0(t) = tP_0(t),\ P_0(t) = 1. \quad (3.27)$$

Now set $2x = 2 + t$ and $P_n(2x - 2) = U_n(x)$. Then, one obtains

$$U_{n+1}(x) + U_{n-1}(x) = 2xU_n(x),\ U_1(x) = 2xU_0(x),\ U_{-1} = 0,\ U_0(x) = 1 \quad (3.28)$$

which, using the Askey classification (see [16]), defines a 3-term recurrence for the Chebyschev polynomials of 2nd kind [16]. From Lemma 2, it follows that the eigenvalues of J^N obtained by truncating the indices $n, m, ...$ as $0 \le n, m, ... \le N-1$ are exactly given by the zeros of the Chebyshev polynomials $U_N(x)$. These latter are given [16] by $z_k^N = \cos(\frac{(k+1)\pi}{N+1})$, $k = 0, 1, ..., N-1$. Therefore, the zeros of $P_N(2x-2)$ are given by $(2z_N^k - 1)$, $k = 0, 1, ..., N-1$. This together with $J := -\frac{1}{\mu^2}G^N$ yields (3.26) and ii) is proven.
Finally, iii) is guaranteed by a theorem due to M. Stone (see in [8]) stating that any self-adjoint operator with simple spectrum is generated by a Jacobi operator of the type considered here. This completes the proof. ■

Remark 3.2 Let us first collect some technical (albeit important) points. We recall that Chebyshev polynomials of second kind are given by

$$U_n(t) := (n+1)_2\mathrm{F}_1(-n,\ n+2;\ \frac{3}{2};\ \frac{1-t}{2}) = \frac{P_n^{\frac{1}{2},\frac{1}{2}}(t)}{P_n^{\frac{1}{2},\frac{1}{2}}(1)},\ \forall n \in \mathbb{N}, \tag{3.29}$$

where ${}_2\mathrm{F}_1$ denotes the hypergeometric function and $P_n^{\alpha,\beta}(x)$ is a particular family of Jacobi polynomials. Recall that for any $N \in \mathbb{N}$, $U_N(t)$ has N different simple roots in $[-1, 1]$. The orthogonality relation among the Chebyshev polynomials U_n is

$$\int_{-1}^{1} d\mu(x)\ U_m(x)U_n(x) = \frac{\pi}{2}\delta_{mn},\ d\mu(x) = dx\sqrt{1-x^2} \tag{3.30}$$

For more details, see e.g [16].
ii) To close this remark, we mention that the Jacobi operator considered here is not completely continuous (although it is of course continuous). Recall that a completely continuous bounded operator maps any weakly convergent sequence into a strongly convergent one. As far as bounded Jacobi operators are concerned, complete continuity is equivalent to the condition that the sequences a_n and b_n go to zero as $n \to \infty$. This condition is not fullfilled here. ♦

Proposition 3.3 *The kinetic operator G on $\ell^2(\mathbb{N})$ (3.25) is positive.*

PROOF For any $v \in \ell^2(\mathbb{N})$, $v = \sum_k v_k e_k \neq 0$, we compute $\langle v, Gv\rangle$. One has:

$$\langle v, Gv\rangle = \mu^2 \sum_{m\in\mathbb{N}} (v_m^\dagger(2v_m - v_{m+1} - v_{m-1})) = \mu^2(|v_0|^2 + \sum_{m\in\mathbb{N}} |(v_m - v_{m+1})|^2) > 0. \tag{3.31}$$

Hence, G is positive. ■

Remark 3.4 i) Since G is positive, there exists a self-adjoint operator D on $\ell^2(\mathbb{N})$ such that $G = D^2$. D may therefore play the role of a Dirac operator, that is bounded since G is bounded. We will examine below the spectral triple built from the standard Moyal triple in the "matrix base" formulation where the usual Dirac operator, says $\partial\!\!\!/$ is replaced by the above operator D.
ii) Proposition 3.3, together with the fact that $G \in \mathcal{B}(\ell^2(\mathbb{N}))$ imply that the spectrum of G, *spec*(G), satisfies *spec*$(G) \subseteq [0, ||G||]$. We now characterize completely the spectrum of G. ♦

We first characterize the point spectrum for the operator G, $spec_P(G)$ then use a general theorem valid for bounded self-adjoint operator to identify the full spectrum. This can be easily performed by simply remarking that the eigenvectors of G^N that gave rise to (3.26) can be used to built ("finite range") eigenvectors for the operator G on $\ell^2(\mathbb{N})$, as a simple corrolary of Lemma 2. Namely,

Lemma 3.5 *i) The point spectrum of the kinetic operator G on $\ell^2(\mathbb{N})$ defined by (3.25) is $spec_P(G) = \cup_{N\in\mathbb{N}^*} spec(G^N)$ where $spec(G^N)$ given by (3.26) with family of related orthonormal eigenvectors of $\ell^2(\mathbb{N})$ given by*

$$v_{N,m} = \sum_{p=0}^{N-1} \left((-1)^m (N+1)\frac{\sin[\frac{N(m+1)\pi}{N+1}]}{\sin^3[\frac{(m+1)\pi}{N+1}]}\right)^{-\frac{1}{2}} U_p(\frac{2+\lambda_{N,m}}{2})e_p, \tag{3.32}$$

where $\{e_p\}_{p\in\mathbb{N}}$ is the canonical basis of $\ell^2(\mathbb{N})$ and corresponding eigenvalues

$$\lambda_{N,m} := 2(\cos(\frac{(m+1)\pi}{N+1}) - 1),\ m \in \{0,1,...,N-1\}, N\in\mathbb{N}^*\}. \tag{3.33}$$

with multiplicity 1.
ii) The spectrum of G is

$$spec(G) = \{\lambda_k^N,\ k\in\{0,1,2,...,N-1\}, N\in\mathbb{N}, N\neq 0\} \cup \{0\}. \tag{3.34}$$

PROOF First, recall that the point spectrum is determined by the eigenvalues of G. We set

$$G_{ml} = \mu^2 \sum_{p\in\mathbb{N}} \mathcal{R}_{mp}\lambda_p \mathcal{R}^\dagger_{pl},\ \sum_{p\in\mathbb{N}} \mathcal{R}_{mp}\mathcal{R}^\dagger_{pl} = \sum_{p\in\mathbb{N}} \mathcal{R}^\dagger_{mp}\mathcal{R}_{pl} = \delta_{ml},\ \mathcal{R}^\dagger_{mn} = \mathcal{R}_{nm}. \tag{3.35}$$

From this one easily obtains $\mathcal{R}_{m+1}(\rho_q) + \mathcal{R}_{m-1}(\rho_q) = (2+\rho_q)\mathcal{R}_m(\rho_q)$, $\forall m,q\in\mathbb{N}$ (and $\mathcal{R}_{-1}(\rho_q) = 0$), where we have set for convenience $\rho_q = -\lambda_q$, and $\mathcal{R}_m(\rho_q) := \mathcal{R}_{mq}$. It is the $x = \rho_q$ evaluation of

$$\mathcal{R}_{m+1}(x) + \mathcal{R}_{m-1}(x) = (2+x)\mathcal{R}_m(x),\ \forall m\geq 1,\ \mathcal{R}_1(x) = (2+x)\mathcal{R}_0(x) \tag{3.36}$$

with $\mathcal{R}_{-1}(x) = 0$. As we did for Lemma 2.11, we first truncate the operator to $N\times N$ submatrice ($0\leq m,l,...\leq N-1$)and set $J^N_{ml} := (-G^N_{ml})$. Then, we can write

$$J^N \cdot \begin{pmatrix} \mathcal{R}_0(x) \\ \mathcal{R}_1(x) \\ \cdots \\ \cdots \\ \mathcal{R}_{N-1}(x) \end{pmatrix} + \begin{pmatrix} 0 \\ 0 \\ \cdots \\ 0 \\ \mathcal{R}_N(x) \end{pmatrix} = x \begin{pmatrix} \mathcal{R}_0(x) \\ \mathcal{R}_1(x) \\ \cdots \\ \cdots \\ \mathcal{R}_{N-1}(x) \end{pmatrix}, \tag{3.37}$$

where (setting $\mathcal{R}_0(x) = f(x)$) $\mathcal{R}_m(x) = f(x)U_m(\frac{2+x}{2})$, $0\leq m\leq N$. This leads immediately to the proof of ii) of Proposition 3.1. We define $\lambda_{N,k} := 2(\cos(\frac{(k+1)\pi}{N+1})-1)$, $k\in\{0,1,...,N-1\}$, $N\in\mathbb{N}^*$ (see (3.26)). On the other hand, one readily realizes from (3.37) that any vector of $\ell^2(\mathbb{N})$ of the form

$$v_{N,k} = \sum_{p=0}^{N-1} \mathcal{R}_p(\lambda_{N,k})e_p = \sum_{p=0}^{N-1} f(\lambda_{N,k})U_p(\frac{2+\lambda_{N,k}}{2})e_p \tag{3.38}$$

is an eigenvector for the operator $G\in\mathcal{B}(\ell^2(\mathbb{N}))$. Hence the point spectrum of G is given by the spectrum of G^N (3.26). Now from (3.35) and the expression for the $\mathcal{R}_n(x)$'s, we obtain

$$\sum_{p=0}^{N-1} \mathcal{R}_p(\lambda_{N,m})\mathcal{R}_p(\lambda_{N,l}) = f(\lambda_{N,m})f(\lambda_{N,l}) \sum_{p=0}^{N-1} U_p(\frac{2+\lambda_{N,m}}{2})U_p(\frac{2+\lambda_{N,m}}{2}). \tag{3.39}$$

This relation can be calculated by using the Christoffel-Darboux formula of Proposition 2.10. (3.39) automatically vanishes whenever $m \neq l$ since it appears only terms involving $U_N(\frac{2+\lambda_{N,k}}{2})$ which are therefore equal to zero since the $\frac{2+\lambda_{N,k}}{2}$'s are the roots of U_N. When $m = l$, one has, setting $t_k^N = \frac{2+\lambda_{N,k}}{2}$,

$$\sum_{p=0}^{N-1}(U_p(t_m^N))^2 = U'_N(t_m^N)U_{N-1}(t_m^N) = \frac{N+1}{((t_m^N)^2-1)}(T_{N+1}(t_m^N)U_{N-1}(t_m^N)), \tag{3.40}$$

where we used $U'_N(x) = \frac{(N+1)T_{N+1}(x)-xU_N(x)}{x^2-1}$ in which $T_N(x)$ is the N-th order Chebyshev polynomial of 1st kind, $T_N(\cos\theta) = \cos(N\theta)$). Standard calculations combining (3.40) to (3.35) and (3.39) yields finally

$$f(\lambda_{N,m}) = \left((-1)^m(N+1)\frac{\sin[\frac{N(m+1)\pi}{N+1}]}{\sin^3[\frac{(m+1)\pi}{N+1}]}\right)^{-\frac{1}{2}},\ 0 \leq p, m \leq N-1. \tag{3.41}$$

Observes finally that (3.39)-(3.41) yield the construction of unit vectors. Hence, $\{v_k^N\}$ (3.38) is an orthonormal family. This completes the proof for i).
Now, by recalling that the spectrum of a self-adjoint operator T involves only (generalized) eigenvalues (i.e eigenvalues or limit point values $\lambda \in \mathbb{R}$ such that there exists a sequence of unit vectors in $Dom(T)$ $\{f_n\}_{n\in\mathbb{N}}$ such that $\lim_{n\to\infty}(T-\lambda\mathbb{I})f_n = 0$). (See e.g Theorem 2 p.170 of [53]). This together with i) completes the proof of ii). ■

Remark 3.6 i) We will show below that G is invertible (see Lemma 3.7). Then, the orthonormal family $\{v_k^N\}$ (3.38) is an orthonormal basis for $\ell^2(\mathbb{N})$.
ii) It is instructive to make further comments on the full spectrum of G (3.34). We have separated the point (discrete) part of the spectrum from the limit value 0. We now re-examine the point ii) of Lemma 3.5 without using the theorem mentionned in the above proof. Let us first show that G has no continuous spectrum, except possibly 0 that will be examined below. In view of point iii) of Remark 3.4, we show that any real value $\lambda \in]0, ||G||]$, $\lambda \neq \lambda_k^N$, $\forall k \in \{0, 1, 2, ..., N-1\}$, $N \in \mathbb{N}, N \neq 0$ is a regular value for the operator G. Indeed, one has $(G-\lambda\mathbb{I})v_k^N = -(\lambda-\lambda_k^N)v_k^N$, with $|\lambda - \lambda_k^N|$ bounded, independently of N and k. Then, any $v \in \ell^2(\mathbb{N})$ can be written as $v = \sum_{N,k}\alpha_k^N v_k^N$ with $\sum_{N,k}|\alpha_k^N|^2 < +\infty$; thus $(G-\lambda\mathbb{I})v = -\sum_{\alpha_k^N}(\lambda-\lambda_k^N)v_k^N$ with $\sum_{N,k}|\alpha_k^N(\lambda-\lambda_k^N)|^2 < +\infty$ still holds true since stemming from the above boundedeness of $|\lambda-\lambda_k^N|$. Then $(G-\lambda\mathbb{I})v \in \ell^2(\mathbb{N})$. Hence, λ is a regular value. To determine if 0 belongs or not to the continuous spectrum, we simply consider the resolvent operator which for $\lambda = 0$ is simply G^{-1} which is unbounded since 0 is in the spectrum. Hence, one can identify $\{0\}$ with the continuous spectrum which is reduced to a unique point in the present case. ♦

Lemma 3.7 *The inverse of the kinetic operator G as given in Proposition 3.1 is defined by the following unbounded operator on $\ell^2(\mathbb{N})$*

$$P_{ml} = \frac{1}{\pi\mu^2}\int_{-1}^{1}dx\ \sqrt{\frac{1+x}{1-x}}U_m(x)U_l(x). \tag{3.42}$$

PROOF This is a direct application of the discussion given at the end of section 2. For further convenience, it is better to express the propagator with the "standard" Chebyshev polynomials. The obtention of (3.42) is just routine calculation that uses the recurrence for the Chebyshev polynomials U_m. Notice that one can easily check for instance $\sum_{l\in\mathbb{N}}G_{ml}P_{lr} =$

$\frac{1}{\pi}\int_{-1}^{1}dx\sqrt{\frac{1+x}{1-x}}(2\delta_{ml}-\delta_{m,l+1}-\delta_{m,l-1})U_l(x)U_r(x)$. Then, use $U_{n+1}(x)+U_{n-1}(x)=2xU_n(x)$ to put the integral into the form $\sum_{l\in\mathbb{N}}G_{ml}P_{lr}=\frac{2}{\pi}\int_{-1}^{1}dx\sqrt{\frac{1+x}{1-x}}(1-x)U_l(x)U_r(x)$ and finally (3.30) to show that P_{ml} is the right inverse of G. In the same way $\sum_{l\in\mathbb{N}}P_{ml}G_{lr}=\delta_{mr}$. ■

Remark 3.8 i) As an illustration of the discussion at the end of the section 2, it is easily observed that $G_{ml}=\langle e_m,Ge_l\rangle$ (3.25) can be cast into the form

$$G_{ml}=\int_{-4}^{0}t\ (P_m(t)P_l(t)d\mu(t)),\ \ d\mu(t)=-\frac{1}{\pi}dt(-\frac{t}{2}(2+\frac{t}{2}))^{\frac{1}{2}} \tag{3.43}$$

where the polynomials P_n satisfy $P_{n+1}(t)+P_{n-1}(t)-2P_n(t)=tP_n(t)$, $P_1(t)-2P_0(t)=tP_0(t)$, $P_0(t)=1$ (see e.g the proof of Proposition 3.1). This indeed agrees with the Favard Theorem 2.8 and especially the spectral theorem A.1 (recall eqn. (A.1)).
ii) From the above analysis, we define the following family of projectors

$$P_k^N:\ell^2(\mathbb{N})\to\mathcal{E}_k^N:=span\{v_k^N\},\ P_k^N:=\lambda_k^N|v_k^N\rangle\langle v_k^N|,\ k\in\{0,1,...N-1\},N\in\mathbb{N}^*. \tag{3.44}$$

By further combining this spectral family with Propositions 3.1 and 2, one infers

$$G^N=\sum_{n=1}^{N}\sum_{k=0}^{n-1}\lambda_k^nP_k^n, \tag{3.45}$$

and one easily realizes that $\lim_{N\to\infty}G^N=G$ where the convergence holds true for the strong operator topology. Indeed, for any $f\in\ell^2(\mathbb{N})$, $f=\sum_{N,k}f_k^Nv_k^N$, a simple computation yields $||(G^L-G)f||_2^2=\sum_{n=L}^{\infty}\sum_{k=0}^{n-1}|\lambda_k^n|^2|f_k^n|^2\le\max\{|\lambda_k^n|^2\}\sum_{n=L}^{\infty}\sum_{k=0}^{n-1}|f_k^n|^2\le\max\{|\lambda_k^n|^2\}||f||_2^2$. Set $S_N:=\sum_{n=1}^{N}\sum_{k=0}^{n-1}|f_k^n|^2$. Obviously, for any $\varepsilon>0$, one can find $N_0\in\mathbb{N}$ such that for any $n>N_0$, one has $|S_n-||f||_2^2|<\varepsilon$. Now, simply write $|\sum_{n=L}^{N}\sum_{k=0}^{n-1}|f_k^n|^2|=|S_N-S_L|=|S_N-||f||_2^2|-S_L+||f||_2^2|$ and uses the convergence condition for S_N to show that $\lim_{L\to\infty}||(G^L-G)f||_2^2=0$ for any $f\in\ell^2(\mathbb{N})$. Hence $\lim_{N\to\infty}G^N=G$. ♦

Keeping in mind the point i) of Remark 3.4 and the point ii) above, we define the following self-adjoint operator $D\in\mathcal{B}(\ell^2(\mathbb{N}))$ such that $D^2=G$

$$D:=\sum_{n=1}^{\infty}\sum_{k=0}^{n-1}\alpha_k^nP_k^n,\ (\alpha_k^n)^2=\lambda_k^n,\ k\in\{0,1,...N-1\},N\in\mathbb{N}^*, \tag{3.46}$$

with $\ker(D)=\{0\}$, which will be used below as a Dirac operator involved in a spectral triple.

4. Dirac operators and spectral triples

We now examine if a spectral triple built from the above Dirac operator can support additional structures such as regularity or summability properties. We have of course in mind a spectral triple with the square root of the kinetic operator as Dirac operator from which the NCFT action (3.19) would be obtained as a spectral action. This would correspond to the noncommutative geometry underlying the gauge-fixed action (3.19).

Some properties of the related spectral triple do not depends on the explicit expression of D but only on the classes of operators it belongs to. Here, one salient operatorial property of D (3.46) is that it is bounded. We will therefore consider first the general case of an unspecified self-adjoint operator $\mathcal{D}$ with $Dom(\mathcal{D})=\ell^2(\mathbb{N})$ (hence automatically bounded). This together with the fact that the relevant $*$-representation η defined in Section 3.5

$$\eta:\mathcal{M}_\theta\to\mathcal{B}(\ell^2(\mathbb{N})),\ \eta(e_m\otimes e_n)=e_m\otimes e_n^*,\ \forall m,n\in\mathbb{N} \tag{4.1}$$

maps elements of $\mathcal{M}_\theta$ into Hilbert-Schmidt operators on $\ell^2(\mathbb{N})$ will single out a limited class of spectral triples. We will then specialize to the Dirac operator (3.46) when necessary.

We denote as usual by $\mathcal{L}^p(\mathcal{H})$, $p \geq 1$, the p-th Schatten ideals of $\mathcal{B}(\mathcal{H})$ (the Hilbert space $\mathcal{H}$ will be set equal to $\ell^2(\mathbb{N})$ in a while). Schatten norm on $\mathcal{L}^p(\mathcal{H})$ is denoted by $||.||_p$. It is convenient to recall the definition of a spectral triple that will be used in the sequel.

Definition 4.1 A spectral triple is the set of data $(\mathbb{A}, \pi, \mathcal{H}, \mathcal{D})$ where $\mathbb{A}$ is an involutive (C^*-)algebra and $\mathcal{H}$ is a separable Hilbert space carrying a faithfull $*$-representation of $\mathbb{A}$ and $\mathcal{D}$ a self-adjoint densely defined operator on $\mathcal{H}$, non necessarily bounded, such that for any $a \in \mathbb{A}$:
i) $\pi(a)\ \text{Dom}(\mathcal{D}) \subseteq \text{Dom}(\mathcal{D})$,
ii) $[\mathcal{D}, \pi(a)] \in \mathcal{B}(\mathcal{H})$,
iii) for any $z \notin spec(\mathcal{D})$, $\pi(a)R_\mathcal{D}(z)$ is a compact operator on $\mathcal{H}$, where $R_\mathcal{D}$ is the resolvent operator of $\mathcal{D}$. ♦

Definition 4.2 A spectral triple is regular if (in obvious notations) $\pi(\mathbb{A}) \cup [\mathcal{D}, \pi(\mathbb{A})] \subseteq \bigcap_{j\in\mathbb{N}} Dom(\delta^j)$ where $\delta T := [|\mathcal{D}|, T]$ for any $T \in \mathcal{B}(\mathcal{H})$. ♦

Notice that Definition 4.1 corresponds to an odd spectral triple. We will introduce additional simple structures in a while to built even triples.

We now specialize to the case $\mathcal{H} = \ell^2(\mathbb{N})$ and $\pi = \eta$ (4.1) but still do not focus on some specific Dirac operator, only assuming that it is self-adjoint with $Dom(\mathcal{D}) = \ell^2(\mathbb{N})$. We can characterize interesting properties of the spectral triple built from $\mathcal{D}$ by the following small theorem. To simplify the notation, we now set $\mathcal{L}^p(\ell^2(\mathbb{N})) = \mathcal{L}^p$.

Lemma 4.3 *The data* $\mathfrak{X}_\mathcal{D} := (\mathcal{M}_\theta, \eta, \ell^2(\mathbb{N}), \mathcal{D}))$ *is a regular spectral triple.*

Proof To prove that $\mathfrak{X}_\mathcal{D}$ is a spectral triple, we first need to use simple properties stemming from the fact that $\mathcal{D}$ is bounded and $\eta(a)$ is Hilbert-Schmidt, $\eta(a) \in \mathcal{L}^2$. Let $R_\mathcal{D}(z)$ denotes the resolvent operator for $\mathcal{D}$. In fact, for any $a \in \mathcal{M}_\theta$ and $z \notin spec(\mathcal{D})$, the following properties hold true: i) $[\mathcal{D}, \eta(a)] \in \mathcal{L}^2$, ii) $\eta(a)R_\mathcal{D}(z) \in \mathcal{L}^2$, iii) $[\eta(a), R_\mathcal{D}(z)] \in \mathcal{L}^2$. Points i), ii) and iii) are direct consequences of the fact that $\mathcal{L}^p$ is a 2-sided ideal of $\mathcal{B}(\ell^2(\mathbb{N}))$. Anyway, it is instructive to give a less direct proof by using Hölder estimates. Namely, use $||T|| \leq ||T||_2$ and $||ATB||_p \leq ||A||||T||_p||B||$ for any $T \in \mathcal{L}^p$, $A, B \in \mathcal{B}(\ell^2(\mathbb{N}))$ to obtain for any $a \in \mathcal{N}_\theta$

$$||[\mathcal{D}, \eta(a)]|| \leq ||[\mathcal{D}, \eta(a)]||_2 \leq 2||\mathcal{D}||||\eta(a)||_2 < +\infty. \tag{4.2}$$

Hence i) is proven. Next, for any $a \in \mathcal{M}_\theta$ and any $z \notin spec(\mathcal{D})$, the following estimate holds true

$$||\eta(a)R_\mathcal{D}(z)||_2 \leq ||\eta(a)||_2||R_\mathcal{D}(z)|| < \infty, ||[\eta(a), R_\mathcal{D}(z)]||_2 \leq 2||R_\mathcal{D}(z)||||\eta(a)||_2 < \infty \tag{4.3}$$

so that $\eta(a)R_\mathcal{D}(z) \in \mathcal{L}^2$, $[\eta(a), R_\mathcal{D}(z)] \in \mathcal{L}^2$ and ii) and iii) are true.
Now, $\mathcal{M}_\theta$ is non unital involutive, $\mathcal{D}$ self-adjoint defined everywhere on $\ell^2(\mathbb{N})$ by assumption and the $*$-representation η (4.1) is faithfull by construction. Besides, property i) implies that $[\mathcal{D}, \eta(a)] \in \mathcal{B}(\ell^2(\mathbb{N}))$ while property ii) implies that for any $z \notin spec(\mathcal{D})$, $\eta(a)R_\mathcal{D}(z)$ is a compact operator on $\ell^2(\mathbb{N})$ and one has obviously $\eta(a)v \in \ell^2(\mathbb{N})$ for any $a \in \mathcal{M}_\theta$. Hence, $\mathfrak{X}_\mathcal{D}$ is a spectral triple.
Let $\delta := [|\mathcal{D}|, .] \in Der(\mathcal{B}(\ell^2(\mathbb{N})))$. Since $|\mathcal{D}| \in \mathcal{B}(\ell^2(\mathbb{N}))$, one has $Dom(\delta) = \mathcal{B}(\ell^2(\mathbb{N}))$ and by simple induction $Dom(\delta^j) = \mathcal{B}(\ell^2(\mathbb{N}))$ for any $j \in \mathbb{N}$. Therefore

$$\bigcap_{j\in\mathbb{N}} Dom(\delta^j) = \mathcal{B}(\ell^2(\mathbb{N})). \tag{4.4}$$

■

But one has $\eta(\mathcal{M}_\theta) \subseteq \mathcal{L}^2$ and $[\mathcal{D}, \pi(\mathcal{M}_\theta)] \subseteq \mathcal{B}(\ell^2(\mathbb{N}))$ so that Definition 4.2 is verified. Hence, the spectral triple $\mathfrak{X}_\mathcal{D}$ is regular.

At this point, two remarks are in order

Remark 4.4 Since $\mathcal{D}$ is bounded together with the fact that we use the natural Hilbert-Schmidt action of $\mathcal{M}_\theta$ on $\ell^2(\mathbb{N})$ throught the representation η, it can be expected that the above spectral triple $\mathfrak{X}_\mathcal{D}$ is, informally speaking, "close to" a Fredholm module. Recall that a Fredholm module, i.e an analytic K-cycle, is defined as the set of data $(\mathbb{A}, \pi, \mathcal{H}, F)$ as in Definition 4.1 with π a $*$-representation and the operator $F \in \mathcal{B}(\mathcal{H})$ is such that for any $a \in \mathbb{A}$, $\pi(a)(F^2-1)$, $\pi(a)(F-F^*)$ and $[\pi(a), F]$ are compact operators on $\mathcal{H}$. Furthermore, a Fredholm module is p-summable if the $*-$algebra $\mathfrak{A} := \{a \in \mathbb{A} : [\pi(a), F] \in \mathcal{L}^p(\mathcal{H}),\ \pi(a)(F-F^*) \in \mathcal{L}^{\frac{p}{2}}(\mathcal{H}),\ \pi(a)(F^2-1) \in \mathcal{L}^{\frac{p}{2}}(\mathcal{H})\}$ is norm dense in $\mathbb{A}$, for non zero $p \in \mathbb{N}$. In the present case, it can be easily realized that the data $\mathcal{F} := (\mathcal{M}_\theta, \eta, \ell^2(\mathbb{N}), F)$ with $F := \frac{\mathcal{D}}{|\mathcal{D}|}$ together with $F := 0$ on $\ker(\mathcal{D})$ and $\mathcal{D}$ (assuming $\mathcal{D}$ is invertible) define a Fredholm module which is 2-summable. ♦

Remark 4.5 From now on, we focus on the Dirac operator (3.46). For further use, we define the isometry

$$\mathfrak{J} : \ell^2(\mathbb{N}) \to \ell^2(\mathbb{N}),\ \mathfrak{J} := \sum_{n,k} u_k^n P_k^n,\ (u_k^n)^2 = 1,\ \forall k \in \{0, 1, ..., n-1\},\ n \in \mathbb{N}^*, \tag{4.5}$$

where the projectors P_k^n are given by (3.44). $\mathfrak{J}$ satisfies

$$\mathfrak{J}^2 = \mathbb{I},\ D\mathfrak{J} = \mathfrak{J}D. \tag{4.6}$$

According to the analysis of Section 3, any arbitrary element of $\mathcal{N}_\theta$ can be written as

$$a = \sum a_{n_1k_1;n_2k_2} v_{k_1}^{n_1} \otimes v_{k_2}^{n_2} \tag{4.7}$$

(with $\sum |a_{n_1k_1;n_2k_2}|^2 < \infty$) where the $v_{k_i}^{n_i}$'s define the orthonormal basis of $\ell^2(\mathbb{N})$ determined by Lemma 3.5. Then, it can be easily realized (in view of e.g (4.5)) that any element of $\mathcal{M}_\theta$ of the form

$$a_c = \sum a_{nk} v_k^n \otimes v_k^n \tag{4.8}$$

commutes with D (3.46). This shows that the spectral triple $\mathfrak{X}_D$, theorem 4.3, related to D does not fullfill one of the necessary conditions for a spectral triple to define a metric space with the Connes spectral distance [31]. This is unlike the Moyal metric space and its homothetic descendants which all verify this necessary condition. Recall that the latter metric commutant condition is

$$[D, \eta(a)] = 0 \iff a = \lambda\mathbb{I},\ \lambda \in \mathbb{C}, \tag{4.9}$$

which is not verified here. In other words, $\mathfrak{X}_D$ is not a spectral metric space in the sense of [28, 33]. Despite this lost of metric structure for the Connes spectral distance, it turns out that the spectral triple $\mathfrak{X}_D$ can be enlarged with additional algebraic structures, as we now show. ♦

Let us introduce the Pauli matrices

$$\sigma_1 = \begin{pmatrix} 0 & 1 \\ 1 & 0 \end{pmatrix}, \sigma_2 = \begin{pmatrix} 0 & i \\ -i & 0 \end{pmatrix},\ \sigma_3 = i\sigma_1\sigma_2 = \begin{pmatrix} 1 & 0 \\ 0 & -1 \end{pmatrix}, \tag{4.10}$$

which satisfy $\{\sigma_\mu, \sigma_\nu\} = 2\delta_{\mu\nu}\mathbb{I}_2,\ \mu, \nu = 1, 2$ and define an irreducible representation of the Clifford algebra $\mathbb{C}l(2)$. We also define:

$$\gamma^\mu := \mathbb{I}_2 \otimes \sigma_\mu,\ \gamma^{\mu+2} := \sigma_\mu \otimes \mathbb{I}_2,\ \mu, \nu = 1, 2, \tag{4.11}$$

such that $\{\gamma_\mu, \gamma_\nu\} = 2\delta_{\mu\nu}\mathbb{I}_4$, $\{\gamma_{\mu+2}, \gamma_{\nu+2}\} = 2\delta_{\mu\nu}\mathbb{I}_4$, $\mu, \nu = 1, 2$. Let us define the Hilbert space as $\mathcal{H}_0 := \ell^2(\mathbb{N}) \otimes \mathbb{C}^4$. We will need to define a real structure and a grading to get an even triple. Accordingly, we introduce the self-adjoint grading operator Γ (i.e defining the chirality) and unitary antilinear operator (i.e charge conjugation) $\mathcal{J}$ to implement reality condition. They are defined by:

$$\Gamma := \gamma_1\gamma_2\gamma_3\gamma_4 = -\sigma_3 \otimes \sigma_3, \ \Gamma^2 = \mathbb{I}_4, \tag{4.12}$$

$$\mathcal{J} := \gamma_2\mathcal{C} = (\mathbb{I}_2 \otimes \sigma_2)\mathcal{C}, \tag{4.13}$$

where $\mathcal{C}$ is the complex conjugation. We next pick the following Dirac operator

$$\mathcal{D} := D\gamma_3 = D\sigma_1 \otimes \mathbb{I}_2, \tag{4.14}$$

with D given in (3.46). We quote a useful set of algebraic properties.

Proposition 4.6 *The following properties holds true:*

$$\mathcal{J}^2 = -\mathbb{I}_2, \ \mathcal{J}\mathcal{D} = \mathcal{D}\mathcal{J}, \ \mathcal{J}\Gamma = -\Gamma\mathcal{J}, \tag{4.15}$$

$$\Gamma^2 = 1, \ \mathcal{D}\Gamma = -\Gamma\mathcal{D}. \tag{4.16}$$

PROOF This is routine computation. ■

Now it is not difficult to equip the spectral triple of lemme 4.3 with additional structures. Namely, one has

Theorem 4.7 *The following data:*

$$(\mathcal{M}_\theta, \pi := \eta \otimes \mathbb{I}_4, \mathcal{H}_0 := \ell^2(\mathbb{N}) \otimes \mathbb{C}^4, \mathcal{D} := D\gamma_3; \mathcal{J} := \gamma_2\mathcal{C}, \Gamma := -\sigma_3 \otimes \sigma_3), \tag{4.17}$$

where D is given by (3.46), is a regular spectral. The triple is even, supports a weak real structure defined by $\mathcal{J}$ and Γ with KO-dimension equal to 2 and verifies the commutant condition:
i) $[\pi(a), \mathcal{J}^{-1}\pi(b^\dagger)\mathcal{J}] = 0, \ \forall a, b \in \mathcal{M}_\theta$,
and the first order condition modulo compact operators:
ii) $[[\pi(a), D], \mathcal{J}^{-1}\pi(b^\dagger)\mathcal{J}] \in \mathcal{K}(\mathcal{H}_0), \ \forall a, b \in \mathcal{M}_\theta$.

PROOF The bounded self-adjoint operator $\mathcal{D}$ has obviously its domain $Dom(\mathcal{D}) = \mathcal{H}_0$, while the $*$-representation π inherits faithfullness from η. One has $\pi(a)Dom(\mathcal{D}) \subseteq \mathcal{H}_0 = Dom(\mathcal{D})$ for any $a \in \mathcal{M}_\theta$, thanks to $\eta(a) \in \mathcal{L}^2$ and the diagonal form of $\pi(a) := \eta(a) \otimes \mathbb{I}_4$. Next, one checks that for any $a \in \mathcal{M}_\theta$

$$[\mathcal{D}, \pi(a)] = [D, \eta(a)]\sigma_1 \otimes \mathbb{I}_2, , \tag{4.18}$$

$$\pi(a)(\mathcal{D}^2 + 1)^{-1} = \eta(a)(D^2 + 1)^{-1} \otimes \mathbb{I}_4. \tag{4.19}$$

Then, one has $[D, \eta(a)] \in \mathcal{B}(\ell^2(\mathbb{N}))$ which implies $[\mathcal{D}, \pi(a)] \in \mathcal{B}(\mathcal{H}_0)$ thanks to the action on $\mathcal{H}_0$ of this operator, see (4.18). Indeed, set $\ell(a) := [\mathcal{D}, \pi(a)]$. Then, for any $f \in \mathcal{H}_0$, $\langle \ell(a)f, \ell(a)f\rangle = \langle f, \ell(a)^\dagger\ell(a)f\rangle = \langle f, [D, \eta(a)]^\dagger[D, \eta(a)] \otimes \mathbb{I}_4 f\rangle = \sum_i \langle [D, \eta(a)]f_i, [D, \eta(a)]f_1\rangle$, where $f = (f_i)_{i=1,\dots,4}$ and the RHS of the last equality is bounded because $[D, \eta(a)]$ is a bounded operator.
In the same way, $\eta(a)(D^2 + 1)^{-1} \in \mathcal{L}^2$ implies that $\pi(a)(\mathcal{D}^2 + 1)^{-1} \in \mathcal{L}^2(\mathcal{H}_0)$ (in view of its diagonal action, see (4.19)), hence it is compact. Thus, from the Definition 4.1, $(\mathcal{N}_\theta, \pi := \eta \otimes \mathbb{I}_4, \mathcal{H}_0 := \ell^2(\mathbb{N}) \otimes \mathbb{C}^4, \mathcal{D} := D\gamma_3)$ is a spectral triple.
Next, $|\mathcal{D}|$ is bounded and $\delta := [|\mathcal{D}|, .] \in Der(\mathcal{B}(\mathcal{H}_0))$ has its domain $Dom(\delta) = \mathcal{B}(\mathcal{H}_0)$. As for Lemma 4.3, one has $\cap_{j\in\mathbb{N}} Dom(\delta^j) = \mathcal{B}(\mathcal{H}_0)$. Besides, $\pi(\mathcal{M}_\theta) \subseteq \mathcal{L}^2(\mathcal{H}_0)$ and $[\mathcal{D}, \pi(\mathcal{M}_\theta)] \subseteq$

Journal of Physics: Conference Series **634** (2015) 012006 doi:10.1088/1742-6596/634/1/012006

$\mathcal{B}(\mathcal{H}_0)$. Hence $\pi(\mathcal{M}_\theta)[\mathcal{D}, \pi(\mathcal{M}_\theta)] \subseteq \cap_{j\in\mathbb{N}} Dom(\delta^j)$ and the triple is regular.
The algebraic properties, Proposition 4.6, imply immediately that the triple is even with weak real structure (see Remark 4.8 below) defined by $\mathcal{J}$ and Γ and with KO-dimension equal to 2, which can be read off from e.g the table of 2nd of ref [1] (pp. 192, Definition 1.124).
We have also to consider the conditions of commutant and first order. As mere consequences of $\mathcal{J} \in \mathcal{B}(\mathcal{H}_0)$, $\pi(a) \in \mathcal{L}^2(\mathcal{H}_0)$ and the fact that $\mathcal{L}^2(\mathcal{H}_0) \subset \mathcal{K}(\mathcal{H}_0)$ is a two-sided ideal of $\mathcal{B}(\mathcal{H}_0)$, one has immediately

$$[\pi(a), \mathcal{J}^{-1}\pi(b^\dagger)\mathcal{J}] \in \mathcal{K}(\mathcal{H}_0),\ [[\pi(a), D], \mathcal{J}^{-1}\pi(b^\dagger)\mathcal{J}] \in \mathcal{K}(\mathcal{H}_0),\ \forall a, b \in \mathcal{M}_\theta. \tag{4.20}$$

Therefore, the point ii) of the theorem is verified. Let us compute the LHS of the first relation. One has for any $f = (f_i)_{i=1,...,4} \in \mathcal{H}_0$ and $b \in \mathcal{M}_\theta$, $\mathcal{J}^{-1}\pi(b^\dagger)\mathcal{J}f = (R(b) \otimes \mathbb{I}_4)f$ where R is the right multiplication. But one can also write $\pi(a)f = (L(a) \otimes \mathbb{I}_4)f$ where L is the left multiplication which verifies $[L(a) \otimes \mathbb{I}_4, R(b) \otimes \mathbb{I}_4] = 0$ for any $a, b \in \mathcal{M}_\theta$. Hence the commutant condition i) holds true. ■

Remark 4.8 At this point, some comments are in order:
i) It is worth pointing out that we consider here a weak notion of real structure, as the one used e.g in spectral triples on quantum groups for which commutant condition and first order condition are fullfilled only modulo a 2-sided ideal in compact operators [54].
ii) The Dirac operator used in Theorem 4.7 does not satisfies the Leibnitz rule. This latter , if it would have been satisfied, would have implied that the first order condition is verified, $[[\pi(a), D], \mathcal{J}^{-1}\pi(b^\dagger)\mathcal{J}] = 0$.
iii) Notice that Proposition 4.6 still holds true with $\mathcal{J}$ replaced by $\mathcal{J}' = \gamma_2\mathfrak{J}\mathcal{C}$ where $\mathfrak{J}$ given by (4.5). This provides another weak real structure for the spectral triple of Theorem 4.7 with $\mathcal{J}$ replaced by $\mathcal{J}'$, with the same properties, except the commutant property, point i) of Theorem 4.7, which is verified only modulo compact operators.

5. Discussion

Bounded or unbounded Jacobi operators defined by 3-term recurrence are usually associated with families of hypergeometric orthogonal polynomials pertaining to the Askey scheme. These operators appear as kinetic operators in NCFT when expressed as "matrix models" within some matrix base formalism, according to the usual liturgy of NCFT, provided some law of indice conservation holds true. This is the case of NCFT on Moyal space and Moyal plane, e.g the Grosse-Wulkenhaar model and its extensions and the scalar NCFT and gauge theory models on $\mathbb{R}^3_\lambda$. Thus, the case of kinetic Jacobi operator covers numerous NCFT of interest.

In this paper, we have only considered the case of bounded Jacobi operators for which interesting conclusions can already be drawn. The unbounded case will be considered elsewhere [22]. As a representative example, we have analyzed in detail, using the mathematical toolkit developed in the section 2, a particular noncommutative gauge theory model on the Moyal plane introduced in [26]. It is obtained from the so called induced gauge theory of [23, 24] expanded around a non trivial symmetric vacuum and suitably gauge-fixed. For related details, [26]. By the way, a short review on the present situation of noncommutative gauge theories is given in subsection 3.1. The resulting kinetic operator G is bounded Jacobi, related to Chebyshev polynomials of 2nd order.

A simple application of Section 2 to the above NCFT yields a complete characterization of the (spectral) properties of the kinetic operator, its full spectrum and the corresponding propagator, therefore improving and clarifying the results of [26]. This combined with general considerations permits one to obtain robust conclusions on the expected perturbative behavior of the NCFT as well as on information on its physical properties that we now discuss. First, the UV and IR regions are naturally identified respectively with large and small indices, says $m, n, ... \gg 1$ or $m, n, ... = 0, 1$, by simply rescaling the gauge-fixed action by a factor g^2 with mass dimension $[g] = 1$ (hence $[\mu] = 1$), which then from (3.34) correspond respectively to large or small eigenvalues for the "Laplacian" G. From Lemma 3.5, one sees that G has 0 as a limit point in its spectrum which can be interpreted as a kinetic operator of a "massless" theory. Accordingly, the corresponding propagator Lemma 3.7 is unbounded. At the perturbative level, one thus expect that this NCFT will exhibit - correlations at large separation between indices, which can be realized at the level of the eigenvalues of the propagator which do not decay in the UV region. This is unlike, e.g the Grosse-Wulkenhaar model [9, 10] or the scalar of gauge NCFT on $\mathbb{R}^3_\lambda$ [14] characterized by propagators with sufficient UV damping and suppression of correlations at large indices separation.

Such a behavior for the above gauge model renders a perturbative treatment questionable and simply reflects the general fact that a bounded (kinetic) operator cannot give rise to a compact propagator (i.e inverse) with a decaying behavior (in e.g some UV region). As far as the Askey-scheme is concerned, bounded Jacobi operators correspond to the serie of orthogonal Jacobi polynomials and descendants (Jacobi, Gegenbauer, Chebyshev of type 1 and 2, spherical, Legendre polynomials). Any related NCFT and/or matrix model is expected to have a similar problematic perturbative behavior. This however, does not forbid some among these models to be solvable.

The cubic vertex in the gauge-fixed action triggers the appearance at the one-loop order of a non-vanishing one-point (tadpole) function. The corresponding computation is given in the Appendix B for the sake of completeness. This is already the case for the gauge theory on $\mathbb{R}^3_\lambda$ considered in [14]. One-loop non-vanishing tadpole signals quantum vacuum instabilities and an interesting question that deserves further investigation is to examine if a noncommutative analog of the Goldstone theorem can be defined.

The initial action that gave rise to the NCFT studied here is the so called induced gauge theory action (3.1). Recall that this later can be related to a particular NCG described by a finite volume spectral triple [27] homothetic, as noncommutative metric spaces, to the standard noncommutative Moyal space [28]. However, the NCFT (3.19) is obtained from (3.1) after an expansion around some vacuum and a gauge-fixing which thus modify the initial kinetic operator in (3.1) and so the related Dirac operator. Then, one can expect a change in the related NCG compared to the one described by a finite volume triple. By using a Dirac operator obtained from the kinetic operator, we have shown that one can construct an even, regular spectral triple with weak real structure and KO-dimension equal to 2, obeying the commutant condition and the first order condition only modulo compact operators. This spectral triple however does not define a noncommutative metric space for the Connes spectral distance, unlike the initial NCG.

Acknowledgments: Discussions and correspondences with P. Martinetti, N. Pinamonti, H. Steinacker and P. Vitale are gratefully acknowledged. One of us (J.C.W) thanks M. Dubois-Violette for discussions on mathematical aspects related to the first order conditions and twists in noncommutative geometry.

Appendix A. Definitions and spectral theorem for bounded operators

For details, see e.g [55]. Let $\mathcal{H}_1$ and $\mathcal{H}_2$ be 2 separable Hilbert spaces on $\mathbb{C}$ with inner product $\langle , \rangle_{\mathcal{H}_i}$, $i = 1, 2$. Let $T : \mathcal{H}_1 \to \mathcal{H}_2$ be a *linear* operator. We recall that the operator norm is defined by $||T|| := \sup\{\frac{||Tu||_{\mathcal{H}_2}}{||u||_{\mathcal{H}_1}},\ u \in \mathcal{H}_1\}$. The following usefull definitions that will be used in the paper are collected below.

Definition A.1 T is bounded if $||T|| < \infty$. The adjoint operator of a bounded operator is a linear map $T^* : \mathcal{H}_2 \to \mathcal{H}_1$ satisfying $\langle u, T^*v\rangle_{\mathcal{H}_1} = \langle Tu, v\rangle_{\mathcal{H}_2}$, $\forall u \in \mathcal{H}_1$, $\forall v \in \mathcal{H}_2$. An operator $T : \mathcal{H}_1 \to \mathcal{H}_1$ is self-adjoint if $T^* = T$; it is normal if $TT^* = T^*T$; it is unitary if $TT^* = T^*T = \mathbb{I}_{\mathcal{H}}$. An operator $T : \mathcal{H}_1 \to \mathcal{H}_1$ is symmetric if $\langle Tu, v\rangle = \langle u, Tv\rangle$. A projection is defined as a linear operator $P : \mathcal{H}_1 \to \mathcal{H}_1$ such that $P^2 = P$. ♦

One of the main notions entering the version of the spectral theorem we use in this paper is the resolution of the identity of a Hilbert space. A convenient definition is provided below:

Definition A.2 A resolution of the identity E of a Hilbert space $\mathcal{H}$ is a Borel measure on $\mathbb{R}$, projection valued, satisfying for any Borel set U, V the following properties: i) $E(U)$ is a self-adjoint projection, ii) $E(U \cap V) = E(U)E(V)$, iii) $E(\emptyset) = 0$, $E(\mathbb{R}) = \mathbb{I}_{\mathcal{H}}$, iv) $U \cap V = \emptyset \Rightarrow E(U \cup V) = E(U) + E(V)$, v) for any $u, v \in \mathcal{H}$ a complex Borel measure is defined by the map $U \mapsto E_{u,v} = \langle E(U)u, v\rangle_{\mathcal{H}}$. ♦

Let $\mathcal{B}(\mathcal{H})$ denotes the Banach algebra of bounded operators on $\mathcal{H}$. Recall that the resolvent set of any $T \in \mathcal{B}(\mathcal{H})$, $\rho(T)$, is defined as $\rho(T) := \{z \in \mathbb{C}, R(z) := (T - z\mathbb{I})^{-1} \in \mathcal{B}(\mathcal{H})\}$ where $R(z)$ is the resolvent operator. The spectrum of $T \in \mathcal{B}(\mathcal{H})$ is defined by $spec(T) := \mathbb{C}/\rho(T)$ and is thus the subset of $\mathbb{C}$ for which $(T - z\mathbb{I})$ has no bounded inverse. The spectrum of a bounded operator T is a compact subset of the disk of radius $||T||$ while a bounded self-adjoint operator has real spectrum which is a subset of $[-||T||, ||T||]$.

In this paper, we deal actually with bounded self-adjoint operators. The convenient version of the spectral theorem for these operators[4] that will provide the integration measures ensuring orthogonality among each family of polynomials can be stated as:

Theorem A.1 *Let $T : \mathcal{H} \to \mathcal{H}$ denotes a bounded self-adjoint operator. There exists a unique resolution of the identity on $\mathcal{H}$ E satisfying*

$$\langle Tu, v\rangle_{\mathcal{H}} = \int_{\mathbb{R}} t\ dE_{u,v}(t) \tag{A.1}$$

and which is supported on the spectrum of T, $spec(T)$, verifying $spec(T) \subset [-||T||, ||T||]$. For any Borel set $U \subset \mathbb{R}$, $E(U)$ commutes with T.

The above spectral theorem yields the well known functional calculus for self-adjoint operators. It turns out that the above measure $E_{u,v}$ can be actually computed from the Stieltjes-Perron inversion formula:

Theorem A.2 *Let $T : \mathcal{H} \to \mathcal{H}$ denotes a bounded self-adjoint operator with resolvent operator $R(z)$. For any open set $]a, b[\subset \mathbb{R}$, $u, v \in \mathcal{H}$, one has in the strong operator topology sense:*

$$E_{u,v}(]a, b[) = \lim_{\varepsilon \to 0} \lim_{\nu \to 0} \int_{a+\varepsilon}^{b-\varepsilon} \langle R(x + i\nu)u, v\rangle - \langle R(x - i\nu)u, v\rangle dx. \tag{A.2}$$

[4] General definitions and properties for operators are recalled in the appendix Appendix A.

Appendix B. One-point function computation

Adding to $S[\phi]$ (3.19) a source term $\sum_{m,n}\phi_{mn}J_{nm}$, the perturbative expansion stems from the generating functional of the connected correlation functions $W[J]$. It is defined by

$$\begin{aligned}\mathcal{Z}[J] &= \int \prod_{m,n} d\phi_{mn} e^{-S[\phi]-\sum_{m,n}\phi_{mn}J_{nm}} = e^{W[J]} &\text{(B.1)}\\ W[J] &= \ln \mathcal{Z}(0) + W_0[J] + \ln\left(1 + e^{-W_0[J]}(e^{-S_{int}(\frac{\delta}{\delta J})} - 1)e^{W_0[J]}\right) &\text{(B.2)}\\ W_0[J] &= \frac{1}{2}\sum_{m,n,k,l} J_{mn}P_{mn;kl}J_{kl}. &\text{(B.3)}\end{aligned}$$

$S_{int}[\phi]$ denotes the interaction terms (3.21) with $\Omega^2 = \frac{1}{3}$ and the J_{nm}'s are sources. The propagator $P_{mn;kl}$ is readily obtained from Lemma 3.7 and (3.24) . Expanding the logarithm as a formal series yields all the connected diagrams while the effective action $\Gamma[\phi]$ is obtained from $W[J]$ by Legendre transform

$$\Gamma[\phi] = \sum_{m,n}\phi_{mn}J_{nm} - W[J],\ \phi_{mn} = \frac{\delta W[J]}{\delta J_{nm}}|_{J=0}. \quad \text{(B.4)}$$

The formal expansion of (B.2) gives rise to

$$\begin{aligned}W^1[J] &= \sum i\frac{2}{3}\sqrt{3\mu^2}(\delta_{m+1,n} - \delta_{m,n+1})\big((P_{lm;kl} + P_{kl;lm})(P_{nk;cd}J_{cd} + J_{ab}P_{ab;nk})\\ &+ (P_{lm;nk} + P_{nk;lm})(P_{kl;cd}J_{cd} + J_{ab}P_{ab;kl})\\ &+ (P_{kl;nk} + P_{nk;kl})(P_{lm;cd}J_{cd} + J_{ab}P_{ab;lm})\big). &\text{(B.5)}\end{aligned}$$

By usual Legendre transform applied to (B.5), $J_{mn} = \sum_{k,l} G_{nm;kl}\phi_{kl} + ...$, one obtains finally ($\sigma = i\frac{2}{3}\sqrt{3\mu^2}$):

$$\begin{aligned}\Gamma^1[\phi] &= \sigma \sum_{m,n,k,l} v_{mn}\big(\delta_{mk}(P_{ll} + P_{km})\phi_{kn} + \delta_{nl}(P_{kk} + P_{nl})\phi_{ml}\\ &+ (\delta_{m+l,n+k}P_{lk} + \delta_{n+k,m+l}P_{nm})\phi_{lk}\big)\\ &= \sigma\sum(2P_{ll} - P_{l,l+1} + P_{kk} + P_{k+1,k+1} - P_{k,k+1})(\phi_{k,k+1} - \phi_{k+1,k}). &\text{(B.6)}\end{aligned}$$

[1] Connes A, *Noncommutative Geometry* , Academic Press Inc., San Diego (1994), available at http://www.alainconnes.org/downloads.html. Connes A and Marcolli M, *A walk in the noncommutative garden,* (2006), available at http://www.alainconnes.org/downloads.html.

[2] Gracia-Bondía J M., Várilly J C and Figueroa H, *Elements of Noncommutative Geometry,* Birkhaüser Advanced Texts, Birkhaüser Boston, Basel, Berlin (2001).

[3] Doplicher S, Fredenhagen K and Roberts J E, Space-time quantization induced by classical gravity, Phys. Lett. B**331** (1994) 39.

[4] Witten E, Noncommutative Geometry And String Field Theory, Nucl. Phys. B**268** (1986) 253.

[5] Dubois-Violette M, Kerner R and Madore J, Noncommutative differential geometry of matrix algebras, J. Math. Phys. **31** (1990) 316. Dubois-Violette M, Kerner R and Madore J, Non- commutative differential geometry and new models of gauge theory, J. Math. Phys. **31** (1990) 323.

[6] Madore J, Journ. Math. Phys. 32 (1991) 332; Grosse H and Madore J, Phys. Lett. B**283** (1992) 218.

[7] Douglas M R and Nekrasov N A, Noncommutative field theory, Rev. Mod. Phys. **73**, 977 (2001) [*Preprint* hep-th/0106048]. Szabo R J, Quantum field theory on noncommutative spaces, Phys. Rept. **378**, 207 (2003) [*Preprint* hep-th/0109162]. Wallet J-C, Noncommutative Induced Gauge Theories on Moyal Spaces, J. Phys. :Conf. Ser.**103** (2008) 012007, [*Preprint* arXiv:0708.2471].

[8] Akhiezer N I, *The Classical Moment Problem and Some Related Questions in Analysis,* Hafner 1965.

[9] Grosse H and Wulkenhaar R, Power-counting theorem for non-local matrix models and renormalisation, Commun. Math. Phys. **254** (2005) 91. Grosse H and Wulkenhaar R, *Renormalisation of ϕ^4-theory on noncommutative $\mathbb{R}^2$ in the matrix base, J. High Energy Phys.* JHEP**12**(2003)019.

[10] Grosse H and Wulkenhaar R, Renormalisation of ϕ^4-theory on noncommutative $\mathbb{R}^4$ in the matrix base, Commun. Math. Phys. **256** (2005) 305.

[11] de Goursac A and Wallet J-C, Symmetries of noncommutative scalar field theory, J. Phys. A: Math. Theor. **44** (2011) 055401, [*Preprint* arxiv:0911.2645].

[12] Vignes-Tourneret F, Renormalization of the orientable non-commutative Gross-Neveu model," Ann. H. Poincaré **8** (2007) 427. Lakhoua A, Vignes-Tourneret F, Wallet J-C, One-loop beta functions for the orientable non-commutative Gross-Neveu model, Eur. Phys. J. C **52**, 735 (2007), [*Preprint* arXiv:hep-th/0701170].

[13] Vitale P and Wallet J-C, Noncommutative field theories on $\mathbb{R}^3_\lambda$: Toward UV/IR mixing freedom, *J. High Energy Phys.* JHEP**04**(2013)115, [*Preprint* arXiv:1212.5131].

[14] Géré A, Vitale P and Wallet J-C, Quantum gauge theories on noncommutative 3-d space, Phys. Rev. D**90** (2014) 045019 [*Preprint* arXiv:1312.6145] .

[15] Alekseev A Y, Recknagel A and Schomerus V, *Brane dynamics in background fluxes and noncommutative geometry, J. High Energy Phys.* JHEP**05**(2000)010 [*Preprint* hep-th/0003187].

[16] Koekoek R, Lesky P A and Swarttouw R F,*Hypergeometric orthogonal polynomials and their q-analogues*, Springer Monographs in Mathematics, Berlin, New York (2010).

[17] Favard J, Sur les polynômes de Tchebicheff , C. R. Acad. Sci., Paris **200**, 2052 (1935).

[18] Stieltjes T J,Recherches sur les fractions continues , Annales de la faculté des Sciences de Toulouse, **8**, J.1-122 (1894), **9** (1895) A.1-47.

[19] Szegö G, *Orthogonal polynomials*, 3rd edition, American Mathematical Society, (1974).

[20] Simon B, The classical moment problem as a self-adjoint finite difference operator, Adv. Math. **137** (1998) 82.

[21] Teschl G, *Jacobi operators and completely integrable nonlinear lattices*, Math. Surveys Monographs **72**, AMS, 2000.

[22] Wallet J-C, Spectral theorem in noncommutative field theories: Unbounded Jacobi operators, in preparation.

[23] de Goursac A, Wallet J-C and Wulkenhaar R, Noncommutative induced gauge theory, Eur. Phys. J. C **51**, 977 (2007), [*Preprint* arXiv:hep-th/0703075].

[24] Grosse H and Wohlgenannt M, Induced Gauge Theory on a Noncommutative Space, Eur. Phys. J. C **52**, 435 (2007), [*Preprint* arXiv:hep-th/0703169].

[25] de Goursac A, Wallet J-C and Wulkenhaar R, On the vacuum states for noncommutative gauge theory, Eur. Phys. J. C **56** (2008) 293, [*Preprint* arxiv:0803.3035]. See also, de Goursac A, Tanasa A and Wallet J-C, Vacuum configurations for renormalizable non-commutative scalar models, Eur. Phys. J. C **53** (2008) 459, [*Preprint* arXiv:0709.3950].

[26] Martinetti P, Vitale P and Wallet J-C, Noncommutative gauge theories on $\mathbb{R}^2_\theta$ as matrix models, *J. High Energy Phys.* JHEP**09**(2013)051, [*Preprint* arXiv:1303.7185].

[27] Grosse H and Wulkenhaar R, 8D-spectral triple on 4D-Moyal space and the vacuum of noncommutative gauge theory , J. Geom. Phys. **62** (2012) 1583. Gayral V and Wulkenhaar R, Spectral geometry of the Moyal plane with harmonic propagation, [*Preprint* arxiv:1108.2184].

[28] Wallet J-C, Connes distance by examples: Homothetic spectral metric spaces, Rev. Math. Phys. **24** (2012) 1250027, [*Preprint* arxiv:1112.3285].

[29] Gayral V, Gracia-Bondia J M, Iochum B, Schücker T, Varilly J C, Moyal Planes are Spectral Triples, Commun. Math. Phys. **246** (2004) 569.

[30] Cagnache E, Wallet J-C, Spectral Distances: Results for Moyal plane and Noncommutative Torus, SIGMA **6** (2010) 026, [*Preprint* arxiv:0912.4185]. Cagnache E, D'Andrea F, Martinetti P, Wallet J-C, The spectral distance on the Moyal plane, J. Geom. Phys. **61** (2011) 1881, [*Preprint* arxiv:0912.0906].

[31] Rieffel M A, Metric on state spaces, Doc. Math. **4** (1999) 559. M.A. Rieffel, Compact quantum metric spaces, in: Operator Algebras, Quantization, and Noncommutative Geometry, Contemp. Math., vol. **365**, AMS, Providence, RI, (2004) 315.

[32] Latremoliere F, Quantum Locally Compact Metric Spaces, J.Funct.Anal. **264** (2013) 362.

[33] Bellissard J, Marcolli M, Reihani K, Dynamical systems on Spectral Metric Spaces , [*Preprint* arxiv:1008.4617 (2010)].

[34] Chamseddine A H, Connes A, van Suijlekom W, Inner Fluctuations in Noncommutative Geometry without the first order condition , J. Geom. Phys. **73** (2013) 222.

[35] ChamseddineA H, Connes A, van Suijlekom W, Beyond the Spectral Standard Model: Emergence of Pati-Salam Unification , *J. High Energy Phys.* JHEP**11**(2013) 132 . Devastato A, Lizzi F and Martinetti P, Grand Symmetry, Spectral Action and the Higgs mass, *J. High Energy Phys.* JHEP **01**(2014)042.

Devastato A and Martinetti P, Twisted spectral triple for the standard and spontaneous breaking of the grand symmetry, *Preprint* arXiv 1411.1320 [hep-th].

[36] Dubois-Violette M, Dérivations et calcul différentiel non commutatif, C. R. Acad. Sci. Paris, Sér. I Math. **307** (1988) 403.

[37] Dubois-Violette M, Lectures on graded differential algebras and noncommutative geometry, In *Noncommutative differential geometry and its applications to physics* (Shonan, 1999), Math. Phys. Stud. 23, Kluwer Acad. Publ., Dordrecht 2001, 245.

[38] Wallet J-C, Derivations of the Moyal algebra and Noncommutative gauge theories, SIGMA **5** (2009) 013, [*Preprint* arXiv:0811.3850].

[39] Cagnache E, Masson T and Wallet J-C, Noncommutative Yang-Mills-Higgs actions from derivation based differential calculus, J. Noncommut. Geom. **5** (2011) 39, [*Preprint* arXiv:0804.3061].

[40] de Goursac A, Masson T and Wallet J-C, Noncommutative ε-graded connections, J. Noncommut. Geom. **6** (2012) 343, [*Preprint* arXiv:0811.3567].

[41] Minwalla S, Van Raamsdonk M and Seiberg N, Noncommutative perturbative dynamics, *J. High Energy Phys.* JHEP**02**(2000)020 [*Preprint* arXiv:hep-th/9912072]; Matusis A, Susskind L and Toumbas N, The IR/UV connection in the non-commutative gauge theories, *J. High Energy Phys.* JHEP**12**(2000)002 [*Preprint* arXiv:hep-th/0002075]. For a possible link between noncommutative field theories and (global description of) the Quantum Hall Effect, see e.g Hellerman S and Van Raamsdonck M, *J. High Energy Phys.* JHEP**10**(2001)039; for older related works, see e.g Georgelin Y and Wallet J-C, Phys. Lett. A**224** (1997) 303 and references therein.

[42] Grosse H and Wulkenhaar R, Self-dual noncommutative φ^4-theory in four dimensions is a non-perturbatively solvable and non-trivial quantum field theory , [*Preprint* arXiv:1205.0465] (2012). Grosse H and Wulkenhaar R, Construction of the ϕ^4_4-quantum field theory on noncommutative Moyal space, [*Preprint* arXiv:1402.1041 (2014)].

[43] Blaschke D N, Rofner A, Sedmik R I and Wohlgenannt M, On Non-Commutative U(1) Gauge Models and Renormalizability,, J. Phys. A **43** (2010) 425401. Blaschke D N, A New Approach to Non-Commutative U(N) Gauge Fields, Europhys. Lett. **91** (2010) 11001.

[44] Blaschke D N, Grosse H and Wallet J-C,Slavnov-Taylor identities, non-commutative gauge theories and infrared divergences, *J. High Energy Phys.* JHEP**06**(2013)038 [*Preprint* arXiv:1302.2903].

[45] Aoki H, Ishibashi N, Iso S, Kawai H, Kitazawa Y, Tada T,Noncommutative Yang-Mills in IIB Matrix Model, Nucl.Phys. B**565** (2000) 176-192

[46] Steinacker H,Emergent Geometry and Gravity from Matrix Models: an Introduction, Class.Quant.Grav. **27**:133001 (2010). Steinacker H, Non-commutative geometry and matrix models, arXiv:1109.5521 (2011).

[47] Grosse H, Lizzi F, Steinacker H, Noncommutative gauge theory and symmetry breaking in matrix models, Phys. Rev. D**81**, 085034 (2010).

[48] Steinacker H, Quantized Gauge Theory on the Fuzzy Sphere as Random Matrix Model , Nucl.Phys. B**679** (2004) 66. Steinacker H, Szabo R J, Localization for Yang-Mills Theory on the Fuzzy Sphere , Commun. Math. Phys. **278** (2008) 193.

[49] Grosse H, Steinacker H, Wohlgenannt M, Emergent Gravity, Matrix Models and UV/IR Mixing , *J. High Energy Phys.* JHEP**04**(2008)023. Blaschke D N, Steinacker H, Wohlgenannt M,Heat kernel expansion and induced action for the matrix model Dirac operator , *J. High Energy Phys.* JHEP**03**(2011)002.

[50] Wallet J-C, Algebraic setup for the gauge fixing of BF and super BF systems, Phys. Lett. B **235** (1990) 71. Stora R, Thuillier F and Wallet J-C, Algebraic structure of cohomological field theory models and equivariant cohomology, in *Infinite dimensional geometry, non commutative geometry, operator algebras, fundamental interactions*, p.266-297, Cambridge Press (1995).

[51] see Blaschke D N, Steinacker H, On the 1-loop effective action for the IKKT model and non-commutative branes, *J. High Energy Phys.* JHEP**10**(2011)120 and references therein.

[52] Gracia-Bondía J M and Várilly J C,Algebras of distributions suitable for phase space quantum mechanics. I,, J. Math. Phys. **29** (1988) 869. Várilly J C and Gracia-Bondía J M,Algebras of distributions suitable for phase-space quantum mechanics. II. Topologies on the Moyal algebra,, J. Math. Phys. **29** (1988) 880.

[53] Helmberg G, "*Introduction to spectral theory in Hilbert space* , North-Holland publishing company, Amsterdam, London (1969).

[54] See e.g Dabrowski L, Landi G, Sitarz A, van Suijlekom W, Varilly J C, The Dirac operator on $SU_q(2)$, Commun.Math.Phys. **259** (2005) 729. Dabrowski L, D'Andrea F, Landi G, Wagner E, Dirac operators on all Podles quantum spheres, J. Noncommut. Geom. **1** (2007) 213.

[55] Dunford N and Schwartz J T, *Linear Operators II: Spectral Theory*, Interscience 1963.

Covariant quantum mechanics applied to noncommutative geometry

Valerio ASTUTI

Dipartimento di Fisica, Università di Roma "La Sapienza", P.le A. Moro 2, 00185 Roma, Italy; INFN, Sez. Roma1, P.le A. Moro 2, 00185 Roma, Italy

E-mail: valerio.astuti@roma1.infn.it

Abstract. We here report a result obtained in collaboration with Giovanni Amelino-Camelia, first shown in the paper [1]. Applying the manifestly covariant formalism of quantum mechanics to the much studied Snyder spacetime [2] we show how it is trivial in every physical observables, this meaning that every measure in this spacetime gives the same results that would be obtained in the flat Minkowski spacetime.

1. Introduction

One of the main approaches to the quantum gravity problem - the one I am interested in describing here - postulates the quantum nature of spacetime as codified in non-trivial commutation relations between spacetime coordinates:

$$[x^\mu, x^\nu] = i\ell F^{\mu\nu}(x, p)$$

This kind of assumption was introduced in the beginning to try to avoid the ultraviolet singularities of field theories - the length scale ℓ seems to provide a natural cutoff, which given the nature of the function $F^{\mu\nu}$ can be covariant or at least the cause of a deformation of the Lorentz group instead of a breaking. Apart for the usefulness in terms of convergence in quantum field theories (problem which was, anyway, approached with renormalization theory methods afterwards) the assumption of coordinates being quantum non-commutative operators has another reason: it can be considered a useful effective theory approach to more fundamental quantum gravity manifestations on spacetime [3, 4]. Theories based on this approach have the strong advantage of being just a deformation of standard, pre-quantum gravity theories. In this way it is easier to keep check of the physics at work, simply evaluating the difference in prediction with respect to the "classical" theories. This should be true for any quantum gravity theory, but in other approach the classical limit, in which standard quantum mechanics or general relativity are recovered, proved not trivial at all. One of the basic ingredients in the analysis used here is the duality between the so called "Kinematical" and "Physical" Hilbert spaces. The first one is the space of the general degrees of freedom of the theories, without connection to the dynamics of the particular system; the second one is the restriction of the kinematical degrees of freedom to the ones allowed by the dynamics of the system under study. These distinction proves crucial in the case of Snyder spacetime, in which it has always been claimed particle position would be quantized, with dicrete spectrum. In all the previous approaches however the distinction between kinematical and physical degrees of freedom was never emphasized. We show that these concerns should not go overlooked in the analysis. The Snyder spacetime [2] is characterized by noncommutativity of coordinates of the form

$$[x^\mu, x^\nu] = i\ell^2 M^{\mu\nu} \tag{1}$$

where ℓ is the length scale of Snyder noncommutativity and $M^{\mu\nu}$ denotes the Lorentz generators. This noncommutative Snyder spacetime has been studied extensively, especially in relation to the possibility of using it to model physical pictures in which spacetime coordinates are discretized. At the level of analysis involving the unphysical kinematical Hilbert space Snyder's proposal does indeed involve a discretization of coordinates, but this is only a formal artifact: the discreteness leaves no trace on the observable properties of particles on the physical Hilbert space. We note that a similar result was shown in the classical theory with a different approach [5].

2. Covariant formulation of quantum mechanics

We here take as starting point the fact that the standard formulation of quantum mechanics should be viewed merely as a sort of gauge-fixed version of the more general and more powerful manifestly-covariant formulation of quantum mechanics. This formulation of quantum mechanics has progressed significantly over the last decade (see, *e.g.* Refs. [6, 7, 8]), rendering increasingly clear how spatial and time coordinates there play exactly the same type of role. In the standard case (commutative spacetime) spatial and time coordinates are well-defined operators [8] on a "kinematical Hilbert space", which is the space of square-integrable functions $L^2(\mathbb{R}^4, dq^0dq^1dq^2dq^3) \sim L^2(\mathbb{R}^2, dp^0dp^1dp^2dp^3)$, with canonical commutation relations

$$[q^\mu, p_\nu] = i\delta^\mu_\nu \ , \quad [q^\mu, q^\nu] = 0 \ , \quad [p_\mu, p_\nu] = 0 \, . \tag{2}$$

Here we denote with q^μ the spacetime coordinates, while the p_μ are of course the conjugate momenta.

Time is not an evolution parameter because in the covariant formulation of quantum mechanics there is no "evolution": dynamics is codified in a constraint, just in the same sense familiar for the covariant formulation of classical mechanics (see, *e.g.*, Ref.[9]). For example, for free particles the constraint simply enforces on-shellness, and the "reduced" Hilbert space obtained from the kinematical Hilbert space by enforcing on-shellness is the so-called "physical Hilbert space".

The physical content of the theory resides in the properties of self-adjoint operators on the physical Hilbert space. Through these one recovers the information on physical evolution of systems from the formal setup which involves a pure-constraint Hamiltonian. This will be crucial for our analysis of the physical implications of Snyder noncommutativity, so we devote a few equations to showing some aspects of its implementation (more details in Refs. [6, 7, 8], and Refs. [10, 11] for a recent application to noncommutative spacetimes). For free particles one has that states of the physical Hilbert space must comply with the requirement

$$\mathcal{H}\psi = [p_0^2 - p_1^2 - m^2]\psi = 0$$

A convenient strategy for implementing the constraint is based [8] on introducing a corresponding scalar product on the physical Hilbert space, such that different kinematical states are projected on the same physical state. Specifically one adopts the following scalar product:

$$\langle\phi|\psi\rangle = \int d\mu(p)\ \delta(\mathcal{H})\,\Theta(p_0)\ \overline{\phi}(p)\,\psi(p) \tag{3}$$

where $\Theta(p_0)$ specifies a restriction [8] to positive-energy solutions of the on-shellness constraint.

It is important to stress that the spacetime coordinates q^μ are self-adjoint operators on the kinematical Hilbert space, but they are not self-adjoint operators on the physical Hilbert space, since $[q^\mu, \mathcal{H}] \neq 0$. So the physical properties of the theory cannot be captured by properties of the operators q^μ. They are coded in properties of combinations of the operators q^μ and p_μ which commute with the Hamiltonian constraint. An example of such physical observables are the Newton-Wigner operators

$$A_i = q_i - \frac{p_i}{p_0} q_0 + i\frac{p_i}{2p_0^2} \tag{4}$$

which indeed are self-adjoint on the physical Hilbert space (in particular $[A_i, \mathcal{H}] = 0$).

How is this regaining us a picture of evolution from the "frozen" pure-constraint setup? This is easily seen by observing that, for example, the expectations of the operators p_i and A_i in a state ψ of the physical Hilbert space,

$$< \psi|p_i|\psi >= \bar{p}_i \ , \quad < \psi|A_i|\psi >= a_i \, ,$$

are meaningful physical properties. These are just expectations of observables in a given (frozen) state that satisfies the Hamiltonian constraint, but contain information on the evolution of the system. One can easily see this by contemplating the interpretation of the Newton-Wigner operator in classical mechanics: the state ψ is placing the particle on a worldline (a fuzzy worldline in quantum mechanics) which in the classical limit is such that

$$q_i = a_i + \frac{\bar{p}_i}{p_0(\bar{p}_i)} q_0$$

(because of the Hamiltonian constraint, on the physical Hilbert space p_0 must be viewed as constrained in terms of the p_i). So the particle is moving with standard velocity $\bar{p}_i/p_0(\bar{p}_i)$ and intercepts the q^i axis at $q^i = a^i$.

3. Discreteness of Snyder spacetime

We reminded our readers in the previous section of the fact that even the standard (commutative) spacetime coordinates are not self-adjoint operators on the physical Hilbert space. They are not observables. They are however well-defined operators on the kinematical Hilbert space, well suited for characterizing the geometric structure of "empty" Snyder spacetime (there are evidently no physical particle on the kineamtical Hilbert space). Then any proposal of noncommutativity of spacetime coordinates must be formulated first on the kinematical Hilbert space, keeping in mind however that the observable manifestations of coordinate noncommutativity can only be its implications for the properties of self-adjoint operators on the physical Hilbert space. In this section we shall see how the Snyder noncommutativity admits a satisfactory description on a kinematical Hilbert space, which in particular involves a discretization of spatial coordinates.

As anticipated in Snyder's original paper, and more recently established in detail in Ref.[12], the algebraic properties of the coordinate operators implies their spectra to be a quantum lattice of spacing ℓ, while time remains a continuous variable. These results have been established by seeking formal Hilbert-space representations of the Snyder algebra, but the question remained so far concerning the physical interpretation of these results. As already stressed above within the standard formulations of quantum theory there is no room for such a Hilbert space, on which in particular the time coordinate is described by an operator.

To derive the discreteness of spatial coordinates it is sufficient to notice that the spatial coordinates and the angular momenta generators form an $SO(4)$ algebra. This is easily seen by posing

$$L_{ij} = M_{ij} \quad , \quad L_{i4} = \frac{x_i}{\ell} \quad , \quad L_{AB} = -L_{BA} \, ,$$

Journal of Physics: Conference Series **634** (2015) 012007 doi:10.1088/1742-6596/634/1/012007

which indeed, also in light of (1), reproduce the commutation relations for the $SO(4)$ algebra:

$$[L_{AB}, L_{CD}] = i(\delta_{AC}L_{BD} - \delta_{BC}L_{AD} - \delta_{AD}L_{BC} + \delta_{BD}L_{AC})$$

This observation shows that one can find a basis for our kinematical Hilbert space which is built of representations of this algebra. The implications of this are best seen by introducing the six operators

$$A_i = \frac{1}{2}\left(L_i + \frac{x_i}{\ell}\right), \quad B_i = \frac{1}{2}\left(L_i - \frac{x_i}{\ell}\right), \tag{5}$$

with $L_i = \epsilon_{ijk}M^{jk}$; this allows us to decouple the relevant $SO(4)$ algebra in two $SU(2)$ algebras: $[A_i, A_j] = i\epsilon_{ijk}A_k$, $[B_i, B_j] = i\epsilon_{ijk}B_k$ and $[A_i, B_j] = 0$, with $A_iA_i = B_iB_i$.

Then we can have basis states labeled by eigenvalues of A_3, B_3, and the casimir A_iA_i:

$$A_3|j, m_A, m_B\rangle = m_A|j, m_A, m_B\rangle \tag{6}$$

$$B_3|j, m_A, m_B\rangle = m_B|j, m_A, m_B\rangle \tag{7}$$

$$A_iA_i|j, m_A, m_B\rangle = j(j+1)|j, m_A, m_B\rangle \tag{8}$$

On these basis states we obtain for the x_3 coordinate, which can be written as $x_3 = \ell\,(A_3 - B_3)$, that

$$x_3|j, m_A, m_B\rangle = \ell(m_A - m_B)|j, m_A, m_B\rangle \tag{9}$$

The eigenvalues m_A, m_B are integer numbers, so the eigenvalues of the x_3 coordinate form a one dimensional lattice of spacing ℓ. The same reasoning applies of course also to the other spatial coordinates, x_1 and x_2, so we can conclude that the spatial part of the Snyder spacetime is a 3-dimensional quantum lattice (but states in the kinematical Hilbert space which are eigenstates of one coordinate are not eigenstates of the other two coordinates), and infinitely degenerate eigenvalues.

4. Kinematical Hilbert-space representation

Snyder himself gave us a representation of this noncommutative coordinates in terms of a set of commutative ones. We keep the notation p_μ, q^μ. So, in particular, we still have that $[q^\mu, p_\nu] = i\delta^\mu_\nu$. The difference is that now the q^μ will not be interpreted as spacetime coordinates, but only as auxiliary operators useful for the description of Snyder's noncommutative coordinates x^μ. We have:

$$x^\mu = q^\mu - \ell^2 p^\mu (p\cdot q)\,. \tag{10}$$

Indeed we have that

$$\begin{aligned}
[x^\mu, x^\nu] &= [q^\mu - \ell^2 p^\mu(p\cdot q), q^\nu - \ell^2 p^\nu(p\cdot q)] = \\
&= -\ell^2\left([p^\mu(p\cdot q), q^\nu] + [q^\mu, p^\nu(p\cdot q)]\right) + \\
&+ \ell^4[p^\mu(p\cdot q), p^\nu(p\cdot q)] = \\
&= i\ell^2\left(p^\mu q^\nu - p^\nu q^\mu\right) = \\
&= i\ell^2\left(p^\mu x^\nu - p^\nu x^\mu\right) = i\ell^2 M^{\mu\nu}
\end{aligned} \tag{11}$$

It is also easy to verify that in order for the x^μ to be hermitian operators the scalar product on the kinematical Hilbert space must involve the following ℓ-deformed integration measure (p^2 denotes as usual $p_\mu p^\mu$)

$$d\mu(p) = \frac{d^4p}{(1-\ell^2p^2)^{5/2}}\,. \tag{12}$$

The coordinates x^μ are self-adjoint operators in this Hilbert space. We shall soon study their spectra.

It is also easy to verify that this construction (as stressed already in Snyder's original work [2]) preserves Lorentz symmetry. This is essentially due to the fact that Snyder's commutation relations (1) are Lorentz invariant and the measure of integration (12) is Lorentz invariant.

5. Effects of discretization on physical observables

Even if, as we saw, kinematical space coordinates are discretized in Snyder spacetime, we have not yet evaluated any physical coordinate living in this spacetime.

Physical observables can only be introduced on the physical Hilbert space, after enforcing a suitable Hamiltonian constraint. The discretization of Snyder's spacetime could be meaningful only to the extent that it would leave a trace in such observables on the physical Hilbert space. Let us then consider the propagation of particles in Snyder's spacetime. As shown above Snyder's proposal leaves Lorentz symmetry unaffected, so in order to describe propagation of particles we need to enforce the standard Hamiltonian constraint

$$[p_0^2 - p_1^2 - m^2]\psi = 0$$

The kinematical Hilbert space is "reduced" to the physical Hilbert space of states ψ that satisfy this constraint. And we start by noticing that the modification of the scalar product affecting the kinematical Hilbert space trivializes when restricted to states in the physical Hilbert space

$$d\mu(p) = \frac{d^4p}{(1-\ell^2p^2)^{5/2}} \longrightarrow d\mu(p) = \frac{d^4p}{(1-\ell^2m^2)^{5/2}},$$

This apparent deformation of the measure of integration just amounts to multiplication by a constant (the mass m is fixed on the entire physical HIlbert space once and for all). It therefore gets reabsorbed in the normalization of the states, and is completely irrelevant.

This observation about scalar products takes us one step closer to establishing the futility of the results obtained for Snyder's spacetime at the level of the kinematical Hilbert space. The possibility of discretization of the coordinates x_j is not even a meaningful possibility on the physical Hilbert space, since evidently the x_j do not commute with the Hamiltonian constraint operator and therefore cannot be observables on the physical Hilbert space. What one could hope for is for the discretization of the x_j on the kinematical Hilbert space to manifest itself under a different disguise as some property of observables on the physical Hilbert space. But we find that this is not the case. To see this let us start from the observables on the physical Hilbert space for free propagating particles which have been most considered in the relevant literature. For the undeformed theory ($\ell = 0$) the operators one usually considers are [13] the operators $\mathcal{X}^\mu = q^\mu - p^\mu(q\cdot v)/(p\cdot v) + h.c.$ where v is a real-valued, time-like vector that parametrizes these operators that commute with the Hamiltonian constraint. This family of operators of course also includes, for $v^\mu = \delta^{\mu 0}$, the Newton-Wigner operator which we already discussed above.

We notice that for the Snyder spacetime one does get good self-adjoint operators on the physical Hilbert space (commuting with the Hamiltonian-constraint operator) by taking these standard $\mathcal{X}^\mu$ and replacing the q^μ with the Snyder coordinates x^μ:

$$\mathcal{X}^\mu_{[\ell]} = x^\mu - \frac{p^\mu}{p\cdot v} x\cdot v + h.c. \qquad (13)$$

It is therefore meaningful to ask whether the discretization of the x_j on the (unobservable) kinematical Hilbert space leaves any trace on properties of these $\mathcal{X}^\mu_{[\ell]}$ observables on the physical

Hilbert space. The answer is no, as one can easily see by recalling that the Snyder coordinates admit the representation $x^\mu = q^\mu - \ell^2 p^\mu (p \cdot q)$, from which it follows that

$$\begin{aligned}
\mathcal{X}^\mu_{[\ell]} &= x^\mu - \frac{p^\mu}{p \cdot v} x \cdot v + h.c. \\
&= q^\mu - \ell^2 p^\mu (p \cdot q) - \frac{p^\mu}{p \cdot v}(q^\nu - \ell^2 p^\nu (p \cdot q)) v_\nu + h.c. \\
&= q^\mu - \frac{p^\mu}{p \cdot v}(q \cdot v) + h.c. = \mathcal{X}^\mu
\end{aligned} \tag{14}$$

The definition of the $\mathcal{X}^\mu_{[\ell]}$ does involve the Snyder deformation scale ℓ but it turns out that actually they are ℓ independent. The $\mathcal{X}^\mu_{[\ell]}$ on the physical Hilbert space are insensitive to the discretization of the x^μ on the kinematical Hilbert space. They are completely independent of the deformation scale ℓ.

On the basis of results obtained in the previous literature on the manifestly covariant formulation of quantum mechanics the observables $\mathcal{X}^\mu_{[\ell]}$ are particularly significant since they come as close as one can get [13] to the notion of a spacetime coordinate on the physical Hilbert space. A mentioned special cases of the $\mathcal{X}^\mu_{[\ell]}$ are the Newton-Wigner operators, to which then evidently our result (14) applies:

$$\begin{aligned}
A^i_{[\ell]} &= x^i - \frac{p^i}{p^0} x^0 + h.c. = \\
&= q^i - \ell^2 p^i (p \cdot q) - \frac{p^i}{p^0}(q^0 - \ell^2 p^0 (p \cdot q)) + h.c. = \\
&= q^i - \frac{p^i}{p^0} + i \frac{p^i}{2(p^0)^2} = A^i
\end{aligned}$$

6. Conclusions

We showed that Snyder's famous discretization of spatial coordinates is no longer found once one considers only physical degrees of freedom, i.e. observables of the system compatible with the dynamics. One of the most important points of this paper is to emphasize the conceptual difference between kinematical degrees of freedom, living in a sort of unphysical space, and the real observables, which live in a different Hilbert space and thus have properties in principle disconnected from the kinematical ones. We emphasize that to make meaningful physical prediction one should extend his statements to the physical Hilbert space, being the kinematical Hilbert space not physically faithful. Still, it should not be expected that all quantum properties of spacetime introduced on the kinematical Hilbert space disappear on the physical Hilbert space of free particles. What we found in our previous study Ref.[11], concerning the so-called "κ-Minkowski noncommutativity", establishes that in some cases the implications of noncommutativity of coordinates, introduced on the kinematical Hilbert space, does affect significantly the structure of the physical Hilbert space, even with just free particles. So this appears to be an issue that requires a case-by-case analysis.

[1] Amelino-Camelia G and Astuti V, [*Preprint* arXiv:1404.4773v1[hep-th]].
[2] Snyder H S, Phys. Rev. **71** (1947) 38.
[3] Doplicher S, Fredenhagen K and Roberts J E, Commun. Math. Phys. **172** (1995) 187 [*Preprint* hep-th/0303037].
[4] Majid S and Ruegg H, Phys. Lett. B **334** (1994) 348 [*Preprint* hep-th/9405107].
[5] Mignemi S, [*Preprint* arXiv:1308.0673v2[hep-th]]

[6] Halliwell J J, Phys. Rev. D **64** (2001) 044008 [*Preprint* gr-qc/0008046].
[7] Gambini R and Porto R A, Phys. Rev. D **63** (2001) 105014 [*Preprint* gr-qc/0101057].
[8] Reisenberger M and Rovelli C, Phys. Rev. D **65** (2002) 125016 [*Preprint* gr-qc/0111016].
[9] Henneaux M and Teitelboim C, *Quantization of gauge systems*, Princeton, USA: Univ. Pr. (1992) 520 p.
[10] Amelino-Camelia G, Astuti V and Rosati G, Eur. Phys. J. C **73** (2013) 2521 [*Preprint* arXiv:1206.3805 [hep-th]].
[11] Amelino-Camelia G, Astuti V and Rosati G, Phys. Rev. D **87** (2013) 084023 [*Preprint* arXiv:1304.7630 [hep-th]].
[12] Lu L and Stern A, Nucl. Phys. B **854** (2012) 894 [*Preprint* arXiv:1108.1832 [hep-th]].
[13] Freidel L, Girelli F and Livine E R, Phys. Rev. D **75** (2007) 105016 [*Preprint* hep-th/0701113].

CTCQG2014 IOP Publishing
Journal of Physics: Conference Series **634** (2015) 012008 doi:10.1088/1742-6596/634/1/012008

Noncommutative geometry, Grand Symmetry and twisted spectral triple

Agostino Devastato

Dipartimento di Fisica, Università di Napoli Federico II

INFN, Sezione di Napoli, Monte S. Angelo, Via Cintia, 80126 Napoli, Italy

E-mail: agostino.devastato@na.infn.it ; astinodevastato@gmail.com

Abstract. In the noncommutative geometry approach to the standard model we discuss the possibility to derive the extra scalar field sv - initially suggested by particle physicist to stabilize the electroweak vacuum - from a "grand algebra" that contains the usual standard model algebra. We introduce the Connes-Moscovici twisted spectral triples for the Grand Symmetry model, to cure a technical problem, that is the appearance, together with the field sv, of unbounded vectorial terms. The twist makes these terms bounded, and also permits to understand the breaking making the computation of the Higgs mass compatible with the 126 GeV experimental value.

1. Introduction

Noncommutative geometry [1–4] allows to study a large variety of geometrical frameworks from a totally algebraic approach. Particularly, it is very useful in the derivation of models in high energy physics, for example the Yang-Mills gauge theories [5–9]. The mathematical structure of noncommutative geometry is based on three algebraic objects: a C^*-algebra $\mathcal{A}$, a Hilbert space $\mathcal{H}$ and a generalization of the Dirac operator D . These three elements already naturally appear in some elementary applications of quantum field theory, for instance the harmonic oscillator in which the set of physical observables (position, time, energy) is an associative algebra; spinors, describing the system, belong to an Hilbert space, and the Dirac operator $\partial\!\!\!/ = -i\gamma^\mu\partial_\mu$ which determines dynamic. The set of these three elements $(\mathcal{A}, \mathcal{H}, D)$ is named *spectral triple*. If the triple is supplemented with two other ingredients, a *chirality operator* Γ and an anti-unitary operator J called *real structure*, then the spectral triple is said a *graded-real spectral triple* $(\mathcal{A}, \mathcal{H}, D; J, \Gamma)$.

In the current state, gauge theories require a noncommutative geometry structure that is understood to be an *almost commutative* geometry, i.e. the product of continuous geometry, representing space-time, times an internal algebra of finite dimensional matrices. Not many models can be described in this theory because there are a set of seven mathematical constraints to be respected. For example, the *order-zero condition*

$$\left[a, Jb^*J^{-1}\right] = 0 \ , \forall a, b \in \mathcal{A} \tag{1.1}$$

ensuring the same action of the algebra on particles and antiparticles; or the *first-order condition*

$$\left[[D, a] \, , Jb^*J^{-1}\right] = 0 \ , \forall a, b \in \mathcal{A} \tag{1.2}$$

which ensures to obtain a fluctuated Dirac operator of order one in the derivatives. In this geometric framework the spectral action principle [10] enables the description of the full standard model of high energy physics, including the Higgs mechanism, putting it on the same footing geometrical footing as general relativity and making it possible the unification with gravity. Moreover, noncommutative geometry allows to deal with another classical problem of the standard model: the three gauge coupling constants run with energy, and at energies comprised between $10^{13} - 10^{17} GeV$ they are very close, but, in view of present data, they fail to meet at a single point; a first possible way to overcome this problem is to enlarge the Hilbert space, adding new fermions [11]. However it is also possible to deduce, from the spectral action expansion, higher dimensional terms in the Lagrangian, improving the unification, [12] without touching the Hilbert space.

2. The Grand Symmetry

Another useful aspect of noncommutative geometry concerns the prediction of the Higgs mass. This latter, at unification scale, is a function of the other parameters of the theory, especially the Yukawa coupling of fermions y_f and the value of the unified gauge couplings $g = g_3 = g_2 = \frac{5}{3} g_1$. Assuming there is no new physics between the electroweak and the unification scales, i.e. the *big desert hypothesis*, the flow of the Higgs mass under the renormalization group yields a prediction around $170 GeV$, [7]. It is not possible in the noncommutative geometry approach to the standard model as well as in the usual Weinberg-Salam electroweak theory, to predict a Higgs mass near to its experimental value, $m_H \simeq 126 GeV$ [13] without incurring problems of instability. There is, in fact, an instability in the electroweak vacuum which is meta-stable rather than stable (see [14] for the most recent update). This inconsistency strongly suggests that something may be going on. In particular, particle physicists have shown how it is possible to cure this instability, assuming the existence of a new scalar field - usually denoted σ - suitably coupled to the Higgs. In [15] the noncommutative geometry model has been enlarged obtaining the field σ by turning the Majorana coupling constant y_R , in the finite dimensional part of the Dirac operator, into a field:

$$y_R \to y_R \sigma(x) \tag{2.1}$$

However, by definition the bosonic fields in noncommutative geometry are generated by *inner fluctuations* of the Dirac operator, defining the 1-form connection A,

$$A := \sum_i a_i \left[D, b_i\right] \ , \text{with} \, a_i, b_i \in \mathcal{A} \tag{2.2}$$

and, unfortunately, this is not the case for the field σ because of the first-order condition 1.2 on the Majorana part of the Dirac operator, forcing the related one-form to be zero:

$$\left[[D_M, a], Jb^* J^{-1}\right] = 0 \implies [D_M, a] = 0 \implies A_\sigma = 0 \; . \tag{2.3}$$

A possible way to generate spontaneously σ , is to take advantage of the fermion doubling in the Hilbert space $\mathcal{H}$ of the standard model [16], introducing the Grand Symmetry [17], a model based on a larger symmetry than the usual one and mixing gauge and spin degrees of freedom. The spectral triple of the Grand Symmetry is the same of the standard model apart from the algebra. It is the product of the spectral triple $(C^\infty(M), L^2(M,S), \partial\!\!\!/, \gamma^5, \mathcal{J})$ of a compact Riemannian manifold M – where $L^2(M,S)$ is the Hilbert space of square integrable spinors over M, $\partial\!\!\!/$ is the usual Dirac operator, γ^5 the product of the four euclidean gamma matrices and $\mathcal{J}$ the usual charge conjugation operator – by a finite dimensional spectral triple:

$$\begin{aligned}
\mathcal{H}_F &= \mathcal{H}_R \oplus \mathcal{H}_L \oplus \mathcal{H}_R^c \oplus \mathcal{H}_L^c = \mathbb{C}^{32} \\
D_F &= D_0 + D_M = \begin{pmatrix} 0 & \mathcal{M} & \mathcal{M}_R & 0 \\ \mathcal{M}^\dagger & 0 & 0 & 0 \\ \mathcal{M}_R^\dagger & 0 & 0 & \mathcal{M}^* \\ 0 & 0 & \mathcal{M}^T & 0 \end{pmatrix} \\
\gamma_F &= \mathrm{diag}(\mathbb{I}_8, -\mathbb{I}_8, -\mathbb{I}_8, \mathbb{I}_8) \\
J_F &= \begin{pmatrix} 0 & \mathbb{I}_{16} \\ \mathbb{I}_{16} & 0 \end{pmatrix} cc\cdot
\end{aligned} \tag{2.4}$$

the finite Hilbert space $\mathcal{H}_F$ is the usual one of the standard model for one particles family; the matrix $\mathcal{M}$ contains the quarks, leptons and neutrinos Dirac masses with CKM mixing; while the matrix $\mathcal{M}_R$ contains the Majorana neutrinos mass y_R and forms the Majorana part of the Dirac operator, namely D_M.

The choice of the finite algebra $\mathcal{A}_F$ will be more involved. Under natural assumptions on the representation of the algebra, it is shown in [17] that the algebra in the spectral triple of the standard model should be a sub-algebra of $C^\infty(M) \otimes \mathcal{A}_F$ with

$$\mathcal{A}_F = \mathbb{M}_a(\mathbb{H}) \oplus \mathbb{M}_{2a}(\mathbb{C}) \; , \; \text{ with } a \in \mathbb{N}. \tag{2.5}$$

The algebra of the standard model,

$$\mathcal{A}_{sm} := \mathbb{C} \oplus \mathbb{H} \oplus \mathbb{M}_3(\mathbb{C})$$

is obtained from $\mathcal{A}_F$ for $a = 2$, by using in addition to 1.1 and 1.2, another axiom of the theory, namely the grading condition, i.e. every element of the algebra has to commute with the chirality operator, $[a, \Gamma] = 0$.

Differentely, the Grand Symmetry model starts with the "grand algebra" ($a = 4$ in 2.5),

$$\mathcal{A}_G = \mathbb{M}_4(\mathbb{H}) \oplus \mathbb{M}_8(\mathbb{C}) \tag{2.6}$$

and one can generate the field σ by inner fuctuation, satisfying the first-order condition imposed by the majorana part D_M of Dirac operator D. In fact, by using again the grading and the

Journal of Physics: Conference Series **634** (2015) 012008 doi:10.1088/1742-6596/634/1/012008

first-order conditions one has the following reductions:

$$\mathcal{A}_G \xrightarrow{\text{grading}} \mathcal{A}'_G = [\mathbb{M}_2(\mathbb{H})_L \oplus \mathbb{M}_2(\mathbb{H})_R] \oplus [\mathbb{M}_4(\mathbb{C})_r \oplus \mathbb{M}_4(\mathbb{C})_l] \tag{2.7}$$
$$\mathcal{A}'_G \xrightarrow{\text{1st order}} \mathcal{A}''_G = (\mathbb{H}_L \oplus \mathbb{H}'_L \oplus \mathbb{C}_R \oplus \mathbb{C}'_R) \oplus (\mathbb{C}_l \oplus \mathbb{M}_3(\mathbb{C})_l \oplus \mathbb{C}_r \oplus \mathbb{M}_3(\mathbb{C})_r)$$

with three of the four complex algebras identified $\mathbb{C}_R = \mathbb{C}_r = \mathbb{C}_l$. The σ field will be given by the difference of two elements of the remaining complex algebras $\mathbb{C}_r$ and $\mathbb{C}'_R$:

$$\sigma \sim y_R(c_R - c'_R), \quad \text{with}\, c_R \in \mathbb{C}_R \text{ and}\, c'_R \in \mathbb{C}'_R \tag{2.8}$$

The final reduction to the standard model, that is $\mathbb{C}_R = \mathbb{C}'_R$, $\mathbb{H}_L = \mathbb{H}'_L$ and $\mathbb{M}_3(\mathbb{C})_l = \mathbb{M}_3(\mathbb{C})_r$ is obtained by the first-order condition on the free Dirac operator:

$$\not{D} := \not{\partial} \otimes \mathbb{I}_F. \tag{2.9}$$

3. Twisted Grand Symmetry

If, on one hand, the Grand Symmetry has the advantage to generate the field σ in the same way than the other gauge fields, on the other hand it contains two weak points. The first limit is that the breaking to the standard model algebra $\mathcal{A}_{sm}$ is obtained by a mathematical condition, i.e. the first-order condition imposed on the free Dirac operator:

$$\text{1-st order condition on } \not{D}: \ \mathcal{A}''_G \to \mathcal{A}_{sm}\,. \tag{3.1}$$

Of course, it would be preferable if the reduction to the standard model was obtained by a dynamical breaking rather than a mathematical reduction.

The second and more important question concerns the unboundeness of the one-forms connections. Unfortunately, before the last breaking 3.1 not only is the first-order condition not satisfied, but the commutator

$$[\not{D}, a] \ , \text{ with}\, a \in C^\infty(M) \otimes \mathcal{A}_G \tag{3.2}$$

is never bounded. This is problematic both for physics, because the connection 1-form containing the gauge bosons is unbounded, and from a mathematical point of view, because the construction of a Fredholm module over $\mathcal{A}$ and Hochschild character cocycle depends on the boundedness of the commutator [18].

In [19] both the problems have been solved by using a twisted spectral triple $(\mathcal{A}, \mathcal{H}, D; \rho)$, introduced by Connes and Moscovici in [20] to incorporate type III examples, such as those arising from the transverse geometry of codimension one foliations. Rather than requiring the boundedness of the commutator, one asks that there exists a automorphism ρ of $\mathcal{A}$ such that the twisted commutator

$$[\not{D}, a]_\rho := \not{D}a - \rho(a)\not{D} \tag{3.3}$$

is bounded for any $a \in \mathcal{A}_G$. In fact, one can show that for a suitable choice of a subalgebra $\mathcal{B} \subset C^\infty(M) \otimes \mathcal{A}_G$, a twisted fluctuation of $D = \not{D} + D_M$, satisfying the first order condition, generates a field σ - very similiar to the one of [17]. Together with this new scalar field one has also the generation of an additional bounded vector field X_μ, whose potential will lead a breaking mechanism to the standard model.

Explicitly, $\mathcal{B}$ is the sub-algebra $\mathbb{H}^2 \oplus \mathbb{C}^2 \oplus \mathbb{M}_3(\mathbb{C})$ of $\mathcal{A}_G$. Labelling the two copies of the quaternions and complex algebras by the left/right spinorial indices l, r and the left/right internal indices L, R, that is

$$\mathcal{B} = \mathbb{H}_L^l \oplus \mathbb{H}_L^r \oplus \mathbb{C}_R^l \oplus \mathbb{C}_R^r \oplus \mathbb{M}_3(\mathbb{C}) \tag{3.4}$$

the automorphism ρ consists in the exchange of the left/right spinorial indices:

$$\rho : (q_L^l, q_L^r, c_R^l, c_R^r, m) \to (q_L^r, q_L^l, c_R^r, c_R^l, m) \tag{3.5}$$

where $m \in M_3(\mathbb{C})$ while the c's and q's are complex numbers and quaternions belonging to their respective copy of $\mathbb{C}$ and $\mathbb{H}$.

Furthermore, the breaking to the standard model is now spontaneous, as conjectured in [17]. Namely the reduction of the grand algebra $\mathcal{A}_G$ to $\mathcal{A}_{sm}$ is obtained dynamically, as a minimum of the potential coming from the spectral action. The scalar field σ then plays a role similar as the Higgs field in the electroweak symmetry breaking. Precisely, since the standard model algebra $\mathcal{A}_{sm}$ is the subalgebra of $\mathcal{B}$ invariant under the twist, one can naturally introduce as physical degrees of freedom the fields

$$\Delta(X_\mu) := X_\mu - \rho(X_\mu)\ , \ \Delta(\sigma) := (\sigma - \rho(\sigma))\, D_M \tag{3.6}$$

to measure how far the grand symmetry is from the SM. In (3.6) X_μ and σ are respectively the twisted fluctuations of $\not{\partial}$ and D_M. The new potential, coming from the spectral action expansion, has the form

$$V = V_X + V_\sigma + V_{X\sigma} \tag{3.7}$$

where V_X depends only by the fields $\Delta(X_\mu)$, V_σ depends only by the field $\Delta(\sigma)$ and $V_{X\sigma}$ is an interaction term containing both the fields. In [19, sect. 5.6] it is shown how the minimum of this potential is obtained for

$$\begin{aligned} \Delta(X_\mu) &= 0 \\ \Delta(\sigma) &= 0 \end{aligned} \tag{3.8}$$

which means the invariance of X_μ and σ under the twist ρ. This condition is verified for the biggest invariant sub-algebra of $\mathcal{B}$ i.e. $\mathcal{A}_{sm}$.

4. Conclusions

Let us get the conclusions: the main idea of the *twisted Grand Symmetry* model is that the scalar field σ is related to the dynamical breaking of the grand symmetry to the standard model. This idea was already formulated in [17], but it was not fully implemented because, without the twist, the fluctuation of the free Dirac operator by the grand algebra $\mathcal{A}_G$ yielded an operator whose square was a non-minimal Laplacian. Almost simultaneously, a similar idea was implemented in [21], where the bigger symmetry did not come from a bigger algebra, but it followed from relaxing the first order condition. In the twisted case, here presented, we have shown a reduction mechanism to the standard model very similiar to that of [21]. Therefore, it would be interesting to understand the relationship between the twisted fluctuations and the inner fluctuation without the first oder condition. Moreover, the twist ρ is remarkably simple, and its mathematical structure should be studied more in details, in particular how it should be incorporated in the other axioms of noncommutative geometry, for example the orientability condition where the commutator with the Dirac operator has a very important role. Also, the physical meaning of the twist is intriguing: the un-twisted version of $\mathcal{B}$ forces the action of the algebra to be equal on the right and left components of spinors. In this sense the breaking of the grand symmetry to the standard model could be interpreted as a sort of "primordial" chiral symmetry breaking.

Acknowledgments

The author thanks W. van Suijlekom and F. Lizzi for discussions.

[1] Connes A, *Noncommutative Geometry*, Academic Press, 1984.

[2] Landi G, *An Introduction to Noncommutative Spaces and their Geometries, Springer Lecture Notes in Physics 51*, Springer Verlag (Berlin Heidelberg) 1997. arXiv:hep-th/9701078.

[3] Gracia-Bondia J M, Varilly J C, Figueroa H, *Elements of Noncommutative Geometry*, Birkhauser, 2000.

[4] Connes A, Marcolli M, "Noncommutative Geometry, Quantum Fields and Motives", AMS 2007;

[5] Connes A and Lott J, Particle Models And Noncommutative Geometry (expanded Version), Nucl. Phys. Proc. Suppl. **18B** (1991) 29.

[6] Schucker T, Forces from Connes' geometry, Lect. Notes Phys. **659** (2005) 285 *Preprint* [hep-th/0111236].

[7] Chamseddine A H, Connes A and Marcolli M, Gravity and the standard model with neutrino mixing, Adv. Theor. Math. Phys. **11** (2007) 991 *Preprint* [arXiv:hep-th/0610241].

[8] van den Dungen K and van Suijlekom W D, Particle Physics from Almost Commutative Spacetimes, *Preprint* arXiv:1204.0328 [hep-th].

[9] Jureit J H, Krajewski T, Schucker T and Stephan C A, Seesaw and noncommutative geometry, Phys. Lett. B **654** (2007) 127 *Preprint* [arXiv:0801.3731 [hep-th]].

[10] Chamseddine A H and Connes A, The spectral action principle, Commun. Math. Phys. **186**, 731 (1997) *Preprint* [arXiv:hep-th/9606001].

[11] Andrianov A A, Espriu D, Kurkov M A and Lizzi F, Universal Landau Pole, Phys. Rev. Lett. **111** (2013) 011601 *Preprint* [arXiv:1302.4321 [hep-th]].

[12] Devastato A, Lizzi F, Flores C V, and Vassilevich D, Unification of coupling constants, dimension 6 operators and the spectral action, International Journal of Modern Physics A, 0, 0, 1550033.

[13] Beringer et al. (Particle Data Group), Phys. Rev. D86, 010001 (2012).

[14] Buttazzo D, Degrassi G, Giardino P P, Giudice G F, Sala F, and Salvio A, Investigating the near-criticality of the Higgs boson, *Preprint* arXiv:1307.3536 [hep-ph].

[15] Chamseddine A H and Connes A, Resilience of the Spectral Standard Model, *Preprint* arXiv:1208.1030 [hep-ph].

[16] Lizzi F, Mangano G, Miele G and Sparano G, Fermion Hilbert space and fermion doubling in the noncommutative geometry approach to gauge theories, Phys. Rev. D **55** (1997) 6357 *Preprint* [hep-th/9610035].

[17] Devastato A, Lizzi F and Martinetti P, Grand Symmetry, Spectral Action, and the Higgs mass, *Preprint* arXiv:1304.0415 [hep-th].

[18] Connes A and Moscovici H, Cyclic cohomology and Hopf symmetry, *Preprint* math/0002125 [math-oa].

[19] Devastato A and Martinetti P, Twisted spectral triple for the Standard Model and spontaneous breaking of the Grand Symmetry, *Preprint* arXiv:1411.1320 [hep-th].

[20] Connes A and Moscovici H, Type III and spectral triples, Traces in number theory, geometry and quantum fields, Aspects Math. **E38** (2008), no. Friedt. Vieweg, Wiesbaden, 57–71.

[21] Chamseddine, A H, Connes A and van Suijlekom W D, Inner Fluctuations in Noncommutative Geometry without the first order condition, *Preprint* arXiv:1304.7583 [math-ph].

Two roads to noncommutative causality

Fabien Besnard[1]
[1] EPF, Pôle de recherche M.L. Paris, 3 bis, rue Lakanal, Sceaux F-92330
E-mail: fabien.besnard@epf.fr

Abstract. We review the physical motivations and the mathematical results obtained so far in the isocone-based approach to noncommutative causality. We also give a briefer account of the alternative framework of Franco and Eckstein which is based on Lorentzian spectral triples. We compare the two theories on the simple example of the product geometry of the Minkowski plane by the finite noncommutative space with algebra $M_2(\mathbb{C})$.

1. Introduction

The properties of the electromagnetic field led the early 20th century physicists to a dramatic change in their notions of space and time. Yet more upheaval came from Einstein's theory of the gravitational field. Today, known fields other than gravitation are described together in the Standard Model of particle physics. The derivation of this very complex piece of machinery from noncommutative geometric ingredients is a major breakthrough [1]. While it sheds light on some of the most contrived aspects of the Standard Model, such as the Higgs mechanism, it also opens the door to a renewed understanding of the structure of spacetime. However two hard problems remain. The first is the passage from the euclidean signature to the physical lorentzian one, which is still elusive for noncommutative manifolds (in spite of recent progress in some particular cases, [2]). The second is the quantization of the noncommutative scheme [3].

In fact, there is a third difficulty, which we believe is the key to the elucidation of the first two: the physical interpretation of the theory. In noncommutative geometry, a C^*-algebra is interpreted as an algebra of would-be functions on some virtual space. In the case of the spectral Standard Model the algebra $\mathcal{A}$ is composed of true functions on space with values in $\mathbb{C} \oplus \mathbb{H} \oplus M_3(\mathbb{C})$. On a different stance, coordinates are expected to become noncommutative observables in quantum gravity [4]. We think that it is a fruitful point of view to consider the elements of $\mathcal{A}$ already as quantum observables, which take into account the quantum nature of the "inner" space unraveled by the Standard Model, but not yet the quantum nature of the spacetime manifold.

With this in mind, we try a causal approach to the problem of the lorentzian signature. More precisely, we observe that causality is a property which is specific to lorentzian manifolds, and that moreover the causal order relation encapsulates the metric information up to a conformal factor. This is hardly a new observation: it is a the root of the Causal Sets approach to quantum gravity [5]. Thus, we want to characterize, among the general elements of the algebra of noncommutative functions $\mathcal{A}$, those which play the role of causal functions. The subset of these distinguished elements is called an isocone. The isocone program has been started in [6] and is still in its early stages of developments. In particular it does not yet make contact with the Dirac operator, which is the central tool of noncommutative geometry. In other words it is not yet known how to distinguish in the noncommutative case true causality, which comes from a metric, from general order relations.

However there is another theory, started by Franco and Eckstein in [7], which aims at defining causality out of a Dirac operator. Of course this top-down approach depends on a correct formulation of noncommutative geometry in the lorentzian signature in the first place.

In this paper we will review the isocone-based approach to causality, starting with general physical motivations in section 2. This will lead us to a set of four fundamental hypotheses, which we will transform in section 3 into a mathematically more tractable set of axioms that defines isocones in the C^*-algebra category. In section 4 we will review general properties of isocones, and in section 5 some classification results for them. In section 6 we briefly review the Franco-Eckstein theory, and compare the two approaches on a simple example in section 7. We will then conclude with a few remarks.

In all the text "compact" means "compact and Hausdorff" and the order relation used on $\mathbb{R}$ is always the natural one.

2. Time, clocks and causality

We start with a rather philosophical discussion, where "philosophical" is here an elegant substitute for "vague". This vagueness will perhaps be forgiven since the question we want, to address "what is time ?", is certainly one of the most difficult of all. It seems reasonable to look for an answer in our two best theories of natural phaenomena to date, namely General Relativity and Quantum theory. The former theory carefully distinguishes between chronology, duration and causality, and puts strong structural constraints on them. The latter theory says nothing at all: time in quantum theory is an external parameter. So the punchline appears to be this: time, in any of its three facets mentioned above, is structurally defined by GR and has no quantum aspect at all. Note that by "structurally defined" we mean that there is no more explicit definition of time in GR than there is any explicit definition of sets in axiomatic set theory, but that the two theories implicitly define their objects by specifying how they behave. That the definition of time is a structural one is not really surprising, the only alternative being that it is an emergent phenomenon that can be defined in term of something else. However it is unsatisfying that quantum theory should have no word on this matter, and it is rather unlikely that we have here the definitive answer.

There are indeed several arguments in favor of an influence of quantum theory on the question of time, and we refer to [8] for some of them. Here we quote only two. The first and most obvious is that if the gravitational field has to be quantized, then the structural constraints on time observables will have to be modified accordingly. The second is that the so-called clock hypothesis, which is a postulate of GR, states that a clock measures the proper time of an observer who carries it. But clocks being made of quantum fields, like everything else, cannot in principle be properly modelized without the help of quantum theory. In other words there should be quantum observables associated to them. A possible objection to this point of view would be that clocks are necessarily macroscopic objects. Once again this would amount to say that time is an emergent phenomenon.

The characterization of quantum observables associated to measures of duration poses a real challenge, even in the context of quantum mechanics on a non-relativistic background [9]. However, we will argue that if we fall back to the cruder notion of causality, a tentative answer to this problem can be given. Let us now explain more precisely what we propose to do.

From General Relativity we borrow the concept of *causality*. A spacetime M, i.e. a connected time-orientable Lorentzian manifold, is said to be *causal* if the following relation

$$p \preceq q \Leftrightarrow \text{ there exists a causal curve from } p \text{ to } q \tag{1}$$

is a partial order. Not all metrics define causal spacetimes, even those which solve the Einstein's equations can have pathological causal properties. However, there always exists a neighbourhood U of a given event p such that $\preceq$ is a partial order on U. In fact much more is true: one can find a neighbourhood with the best possible causality property, namely global hyperbolicity (see [10], th 2.14).

The causal ordering permits to define *causal functions*, that is functions $f : M \to \mathbb{R}$ such that $x \preceq y \Rightarrow f(x) \leq f(y)$. These functions are classical observables which correspond to measurement devices that we call *generalized clocks*. A generalized clock, if compared to a genuine clock, can arbitrarily change pace or even stop for a while. The only thing it cannot do is run backwards. We now want to associate quantum observables to these generalized clocks. For this we introduce a quite general setting for quantum physics [11].

(i) For a given physical system, there exists a set of states $\mathcal{S}$, a set of observables $\mathcal{O}$, and a duality map $\mathcal{S} \times \mathcal{O} \to \mathbb{R}$, which is separating for $\mathcal{S}$ and $\mathcal{O}$.
(ii) $\mathcal{S}$ is a convex set.
(iii) $\mathcal{O}$ is a real vector space.
(iv) $\mathcal{O}$ is stable under $a \mapsto f(a)$ where f is a real-valued function.

Note that we do not assume for the moment that the duality function gives the expectation value of the observable when the system is in some state: this is a matter of interpretation of the formalism and we will deal with that later on. There are many good reasons to believe in these axioms, and we refer to [11] for justifications. Let us only comment on the last of them, which will be crucial for our purpose. This axiom follows from the following intuitive interpretation: the observable $f(a)$ corresponds to an observation of a followed by the mathematical operation of applying the function f to any result of such an observation.

A neat way of fulfilling all these requirements is to suppose that $\mathcal{O}$ is the self-adjoint part of a C^*-algebra and that $\mathcal{S}$ is its state space, this is the so-called C^*-paradigm. However it is not quite the most general solution: a Jordan algebra would suffice. So we will start by assuming that $\mathcal{O}$ is a Jordan algebra. We postpone to the next section the specification of the type of Jordan algebra that we will use in order to make sense of our hypotheses.

We now call $\mathcal{C}$ the set of *causal observables* corresponding to generalized clocks. Our first hypothesis, justified above, is that $\mathcal{C}$ is a subset of a Jordan algebra.

Hypothesis 0 (Jordan paradigm) $\mathcal{C}$ is a subset of a Jordan algebra $\mathcal{O}$.

In some classical limit $\mathcal{C}$ should correspond to the set of causal functions on M, but it should be stressed that the correspondence is in the direction causal observable $\to$ causal functions, in particular we do not assume the existence of a quantization map which would go in the other direction. In fact it is well-known that quantization maps cannot commute with functional calculus and be linear at the same time [12]. What are the properties of $\mathcal{C}$? Consider $o \in \mathcal{C}$, c a generalized clock corresponding to o and f a non-decreasing real-valued function. Then the device c' which is formed by c followed by a computation of the function f on the result obtained by c is still a generalized clock. Hence $f(o) \in \mathcal{C}$. This is our hypothesis number 1 for the set $\mathcal{C}$.

Hypothesis 1 (functional stability) If $o \in \mathcal{C}$ and f is a real non-decreasing function, then $f(o) \in \mathcal{C}$.

Let us remark here that this implies that the algebra is unital and that $\mathcal{C}$ contains all the real multiples of the unit. Let us dissipate any worry that could arise at this point: it does not mean that our approach only applies to compact spacetimes. If we start with some noncompact spacetime and use the approach advocated here with a unital commutative algebra, we will recover some compactification of the spacetime we started from, depending on the algebra used.

Of course these objects have boundaries. It is indeed a well-known theorem that compact causal spacetimes always have boundaries (or to put it another way, that compact closed causal spacetimes do not exist).

We now introduce a separation hypothesis.

Hypothesis 2 (separation) $\mathcal{C}$ separates the pure states of $\mathcal{O}$.

The origin of this hypothesis is the classical situation. In any sufficiently small given region of spacetime one can introduce causal coordinates, for instance GPS coordinates [13], which obviously separates events (i.e. pure states). If one considers the spacetime manifolds globally, there can be too few causal functions to separate the events (in fact, there can be none at all, except the constants). However the hypothesis will be fulfilled if one assumes strong causality, which despite its name, is a weaker causality condition than global hyperbolicity [10].

Hypothesis 2 can thus be viewed as an exportation to the quantum regime of the idea that causality plays a fundamental role in the process of identification of what we call "different spacetime events". As such it cannot be fully justified, it is an act of faith.

As soon as we have adopted it, this hypothesis grants us a partial ordering on the set $P(\mathcal{O})$ of pure states of $\mathcal{O}$, which is so defined:

$$\forall p_1, p_2 \in P(\mathcal{O}), p_1 \preceq p_2 \Leftrightarrow p_1(c) \leq p_2(c) \text{ for all } c \in \mathcal{C} \tag{2}$$

In the classical situation, if we started from the set of causal functions, we recover in this way the causal order on spacetime events. Mind that this is not a mere tautology: it could well be that the order defined through (2) turn out to be stronger than the causal order (see the definition of toposets in section 4). But as is now familiar, the equality between causal order and the order defined through the causal functions always holds locally and will hold globally under some causality condition [6]. In the present situation we started from a set of observables that we did not know at first to be associated with some partial order. But we know now from (2) that each observable of our set $\mathcal{C}$ is non-decreasing for $\preceq$. In the classical case the causal functions are exactly those which are non-decreasing for $\preceq$, but even there, we may very well start from a smaller set of causal functions which is still large enough to fulfill the hypothesis 1 and 2 and end up with the same causal order. Hence we see that we need a "saturation" hypothesis to ensure that we started from the whole set of causal functions and not a smaller subset. The most natural is the following:

Hypothesis 3 (saturation) Every observable o which have a non-decreasing Gelfand transform with respect to $\preceq$, i.e. $p_1 \preceq p_2 \Rightarrow p_1(o) \leq p_2(o)$, is in $\mathcal{C}$.

As above, the leap leading to this hypothesis cannot be fully justified. However we will give some arguments in the next section that we hope will give some credence to it.

3. From the fundamental hypotheses to a set of working axioms

In the previous section we put forward four hypotheses, which we qualify as fundamental, about the set $\mathcal{C}$ of causal observables. However, they are not very handy from a mathematical point of view. This is particularly true of the saturation hypothesis. In this section, we will see that under some very reasonable conjectures, they can be turned into axioms which are easier to work with.

First, we must specify the kind of Jordan algebra we use. In order to make sense of hypothesis 1 we must have a functional calculus. We choose JB-algebras, which possess continuous functional calculus. By doing so we subrepticely introduce a (noncommutative) topology, and the question of the compatibility with the causal structure arises. We will see in the next section that a very natural answer to this question can be given thanks to Gelfand theory for toposets.

Let us now use the saturation hypothesis. If $c_1, c_2 \in \mathcal{C}$, then the Gelfand transforms $\hat{c}_1 : p \mapsto p(c_1)$ and $\hat{c}_2 : p \mapsto p(c_2)$ defined on pure states is non-decreasing, hence their sum is also. Therefore $c_1 + c_2 \in \mathcal{C}$ by the saturation hypothesis. Similarly, if c_n is a sequence of elements of $\mathcal{C}$ converging in the norm to c, then each $\hat{c}_n$ is non-decreasing, hence the limit is, hence $c \in \mathcal{C}$.

Consider now the following set of axioms on $\mathcal{C}$, which can be derived from the fundamental hypotheses, as we have seen:

(i) $\mathcal{C} \subset \mathcal{O}$, a JB-algebra.

(ii) $\mathcal{C}$ separates the pure states of $\mathcal{O}$.

(iii) $\mathcal{C}$ is stable by non-decreasing continuous functional calculus.

(iv) $\mathcal{C}$ is norm closed.

(v) $\mathcal{C}$ is stable by sum.

Though it is a priori a weaker set of axioms than the fundamental hypotheses, we conjecture that it is in fact equivalent. We call this the *saturation conjecture*. Whatever the status of this conjecture is, we will work in this paper with this maybe weaker set of axioms. In each case we will encounter, it is possible to show that the set $\mathcal{C}$ is indeed saturated, that is, contains all observables with non-decreasing Gelfand transform.

But first we want to strengthen axiom (ii) by asking instead that:

(ii') $\mathcal{C}$ separates the states of $\mathcal{O}$.

We will show below that axiom (ii) can be replaced with axiom (ii') without loss of generality provided the following property is true for $\mathcal{O}$:

> **Strong Stone-Weierstrass property** Let $\mathcal{O}$ be a JB-algebra and B be a subalgebra of $\mathcal{O}$. If B separates $P(\mathcal{O}) \cup \{0\}$. Then $B = \mathcal{O}$.

The strong Stone-Weierstrass conjecture for JB-algebras states that all JB-algebras satisfy this property. This is known to be true if $P(\mathcal{O})$ is closed and also if $\mathcal{O}$ is post-liminal [14]. All algebras considered in this paper will be post-liminal.

Proposition 1 *If $\mathcal{O}$ satisfies the Strong Stone-Weierstrass property, then the hypotheses (i) to (v) imply (ii').*

The proof is easy. Let B be the strong closure of the set of differences $\mathcal{C} - \mathcal{C}$. First notice that $\mathcal{C} - \mathcal{C} = \mathcal{C}^+ - \mathcal{C}^+$, where $\mathcal{C}^+$ denotes the set of positive elements in $\mathcal{C}$. Indeed, we can always rewrite a difference $a - b$ in the form $a + k.1 - (b + k.1)$ where k is a large enough real and 1 is the unit of $\mathcal{O}$. (Remember that $\mathcal{C}$ contains the real multiples of 1, as noticed in the previous section.) Next, consider $a, b \in \mathcal{C}^+$. Then $a + b \in \mathcal{C}^+$. Moreover, if f is the continuous non-decreasing function such that $f(t) = 0$ for $t < 0$ and $f(t) = t^2$ for $t \geq 0$, then $f(a+b) = (a+b)^2$, $f(a) = a^2$ and $f(b) = b^2$ all belong to $\mathcal{C}^+$. Thus the Jordan product $a \circ b = \frac{1}{2}((a+b)^2 - a^2 - b^2) \in B$. By linearity and continuity of the Jordan product we easily see that B is indeed a subalgebra of $\mathcal{O}$. Now by hypothesis B separates $P(\mathcal{O})$, and since it contains a unit it separates 0 from the other pure states. Thus $B = \mathcal{O}$. Since $\overline{\mathcal{C} - \mathcal{C}} = \mathcal{O}$ it is now obvious that $\mathcal{C}$ separates the states of $\mathcal{O}$.

4. Isocones and I^*-algebras

From now on we concentrate on JB-algebras which are the self-adjoint part of C^*-algebras. In this context the structure of the set of causal observables becomes (under the saturation and Stone-Weierstrass conjectures) that of an isocone, the definition of which we recall below:

Definition 1 *Let A be a unital C^*-algebra, $\mathrm{Re}(A)$ be its self-adjoint part. An isocone of A is a non-empty set I such that:*

(i) $I \subset \mathrm{Re}(A)$,

(ii) $\forall \phi : \mathbb{R} \to \mathbb{R}$ continuous and non-decreasing, $a \in I \Rightarrow \phi(a) \in I$,

(iii) $\forall a, b \in I$, $a + b \in I$,

(iv) I is closed,

(v) I separates the states of A,

If I is an isocone of A, the couple (I, A) is called an I^*-algebra.

Isocones were first defined in [6] but with slight differences, so let us make some comments on them to help make the connection.

First, instead of axiom (v) we required that the set of differences $I - I$ be dense in $\mathrm{Re}(A)$. This is seemingly a stronger condition, but in fact they are equivalent. Indeed, suppose I separates $\mathcal{S}(A)$ and let $V = \overline{I - I}$. Note that V is a real vector space by axioms (ii) and (iii). Suppose there exists $a \in \mathrm{Re}(A)$ such that $a \notin V$ and use Hahn-Banach theorem to find a continuous linear form f such that $f(V) = 0$ and $f(a) \neq 0$. We can extend f as an hermitian form on A and by [18] there exist two states s_1, s_2 on A and two nonnegative numbers α_1, α_2 such that $f = \alpha_1 s_1 - \alpha_2 s_2$, and $\|f\| = \alpha_1 + \alpha_2$. Since $1 \in V$, we find that $\alpha_1 = \alpha_2 = \|f\|/2$. Hence $f = k(s_1 - s_2)$ with $k = \|f\|/2$. By hypothesis $f(V) = 0$, hence s_1 and s_2 are equal on I, and since I separates the states, they are equal, which contradicts $f(a) \neq 0$.

Moreover, in [6] we did not put forward axiom (ii), but instead asked that if $a, b \in I$ and $ab = ba$, then $\sup(a, b)$ and $\inf(a, b)$ must also be in I. This axiom was motivated by the hypotheses of the Kakutani-Stone theorem, thanks to which one can prove an analog of the Gelfand-Naimark theorem for commutative I^*-algebras (see below). However, this mathematically motivated axiom was later shown to be equivalent to the more physically meaningful stability under non-decreasing continuous functional calculus in [15].

As alluded to above, the first step of the theory is to prove an analog of the Gelfand-Naimark theorem for commutative I^*-algebras. Let us introduce some terminology in order to state this theorem. Let $(M, \preceq)$ and $(N, \leq)$ be two partially ordered topological spaces. A map $f : M \to N$ which preserves the order relations is said to be *isotone* or to be an *isotony*. The set of continuous isotonies from M to $\mathbb{R}$, where $\mathbb{R}$ is equipped with the natural order and topology, is denoted by $I(M)$. A *toposet* is a topological ordered space which has "enough" isotonies, in the following sense:

$$\forall x, y \in M, x \preceq y \Leftrightarrow (\forall f \in I(M), f(x) \leq f(y)) \tag{3}$$

This is the compatibility condition between topology and order that we were wondering about in the previous section. A toposet is also called a completely separated topological ordered space. Note that if M is compact, it is a toposet if and only if the relation $\preceq$ is closed as a subset of $M \times M$ [16]. A spacetime is a toposet if and only if it is causally simple [17]. It means that it does not contain closed causal curves and that the future and past of any event are closed. It is the second strongest causality condition after global hyperbolicity.

Theorem 1 *Let M be a compact toposet. Then the couple $(I(M), C(M))$ is a commutative I^*-algebra. Conversely, every commutative I^*-algebra (I, A) is canonically isomorphic to one of this form, where M is the set of pure states of A.*

As we explained before, there cannot be any such duality theorem in the locally compact case where one uses the algebra $\mathcal{C}_0(M)$ of continuous functions vanishing at infinity since this does not contain any non-trivial isotone function in general. Hence one is forced to use a preferred compactification in this case.

The proof of theorem 1 can be found in [6]. Let us just recall that the order on $M = P(A)$ is recovered from I by the formula

$$\forall \phi, \psi \in P(A), \phi \preceq_I \psi \Leftrightarrow (\forall a \in I, \phi(a) \leq \psi(a)) \tag{4}$$

which is just (2) adapted to the present situation. As explained in the previous section, this formula makes sense in any I^*-algebra, commutative or not, and we can see now that it endows $P(A)$ with the structure of a toposet.

Before proceeding let us give a few examples of noncommutative I^*-algebras.

The first is trivial: if I is the whole set of self-adjoint elements of A, all axioms are evidently satisfied. This is called the trivial isocone, and it induces on $P(A)$ the equality ordering $\phi \preceq \psi \Leftrightarrow \phi = \psi$.

For the second example we consider $A = M_2(\mathbb{C})$. In this case it is easily seen that any closed convex cone in ReA which contains the constants and has a non-empty interior is an isocone. We can give a geometric interpretation of the order induced by I on $P(A)$ in the following way. First let us identify $P(A)$ with 2-sphere of rank one projections which we call S. Then $K = I \cap S$ is compact, has non-empty interior and is either equal to S or to a geodesically closed subset of a hemisphere. Then if we write $\omega_{[\xi]}$ for the pure vector state defined by the unit vector ξ and $p_{[\xi]}$ for the projection on the line generated by ξ, we have

$$\omega_{[\xi]} \preceq_I \omega_{[\eta]} \Leftrightarrow (\forall x \in K, d(x, p_{[\xi]}) \geq d(x, p_{[\eta]})) \tag{5}$$

where d is the geodesic distance on S. Clearly we can also use the identification of $P(A)$ with $\mathbb{C}P^1$ and the Fubini-Study metric (see [6] for details).

The next example is infinite-dimensional. If we consider any compact toposet M with a regular Borel measure, we can faithfully represent the elements of $\mathcal{C}(M)$ as multiplicative operators on $L^2(M)$. Consider then the set K of compact operators on $L^2(M)$ and $I = I(M) + \mathrm{Re}(K)$, that is the set of perturbations of isotonies of M by self-adjoint compact operators. Then I is an isocone of $A+K$ [15]. In this case the pure state space of A can be written $P(A) = X \coprod Y$ where $X \approx M$ and Y is the set of vector states. Then the order induced by I is the original order on M extended trivially (no further relations on distinct elements) to $P(A)$. The intuitive idea behind this example is that compact operators correspond to infinitesimals, and that a non-decreasing function stays non-decreasing when one adds an infinitesimal to it.

To build more examples, one can use the two following constructions.

Theorem 2 *Let $(P, \preceq)$ be a finite poset and for each $x \in P$ let (I_x, A_x) be an I^*-algebra. We set $I = \bigoplus_{x \in P} I_x$, $A = \bigoplus_{x \in P} A_x$, and we write elements of A in the form $(a_x)_{x \in P}$. We define*

$$\operatorname*{Lex}_{x \in P} I_x = \{a \in I | \forall x, y \in P, x \prec y \Rightarrow \max \sigma(a_x) \leq \min \sigma(a_y)\}$$

where $\sigma(a)$ denotes the spectrum of a. Then $\mathrm{Lex}_{x \in P} I_x$ *is an isocone of A.*

The isocone $L := \mathrm{Lex}_{x \in P} I_x$ induces the lexicographic sum of the orderings $\preceq_x$ defined by I_x on $P(A_x)$. More precisely put, we have $P(A) = \coprod P(A_x)$ and the pure states are ordered by L in the following way: let $\phi \in P(A_x)$ and $\psi \in P(A_y)$ then

$$\phi \preceq_L \psi \Leftrightarrow (x \neq y \text{ and } x \preceq y \text{ in } P) \text{ or } (x = y \text{ and } \phi \preceq_{I_x} \psi \text{ in } P(A_x)) \tag{6}$$

Journal of Physics: Conference Series **634** (2015) 012009 doi:10.1088/1742-6596/634/1/012009

This is why we call L the lexicographic sum over P of the isocones $(I_x)_{x\in P}$. If the order relation on P is trivial, this is just a direct sum. There is a generalization of this construction to infinite posets, but one needs to add conditions on the order and the map $x \mapsto I_x$. To simplify matters we also suppose that the algebra $A_x = A$ is the same for all x.

Theorem 3 *Let $(P, \preceq)$ be a metrizable compact poset and for each $x \in P$ let I_x be an isocone of A. Let $\mathcal{A}$ be the C^*-algebra $\mathcal{C}(P, A)$ and*

$$\operatorname*{Lex}_{x\in P} I_x := \{f \in \mathcal{C}(A,P) | \forall x \in P, f(x) \in I_x \text{ and } \forall x,y \in P, x \prec y \Rightarrow \max \sigma(a_x) \leq \min \sigma(a_y)\}$$

where $\sigma(a)$ denotes the spectrum of a. If $\prec$ is closed and $x \mapsto I_x$ is lower hemicontinuous, then $\mathrm{Lex}_{x\in P} I_x$ is an isocone of $\mathcal{A}$.

Remember that a multi-valued function $\varphi : X \longrightarrow 2^Y$ is said to be *lower hemi-continuous* if and only if the set $\{x \in X | \varphi(x) \cap V \neq \varnothing\}$ is open for all open $V \subset Y$. The condition that $\prec$ is closed as a subset of the product $P \times P$ is equivalent to the following one: for each $x \in P$ there exists a neighbourhood U of x such that no element of $U \setminus \{x\}$ is comparable to x. It is of course satisfied if P is discrete (hence finite by compactness) but this is not necessary. For instance the ordering on $P = [0;1]$ defined by the relations $x \preceq 1$ for all $x \leq 1/2$ and $x \preceq x$ for all $x \in [0;1]$ satisfies it. Note that if $\prec$ is closed and P is compact then $\preceq$ is closed hence P is a toposet.

The other construction which will play an important role is the "pushforward" by surjective $*$-morphisms.

Proposition 2 *Let A, B be C^*-algebras, $\pi : A \to B$ a surjective $*$-morphism, I an isocone of A. Then $\pi(I)$ is an isocone of B.*

Note that $\pi(I)$ is automatically closed if B is finite-dimensional.

5. Isocones in finite-dimensional and almost-commutative algebras

The first task is to classify isocones in matrix algebras. We will say that an algebra is *egalitarian* if it contains only the trivial isocone (the justification for the terminology is that the only isocone-induced order on its pure state space will be equality).

In [15] we proved the following:

Theorem 4 *The matrix algebra $M_n(\mathbb{C})$ is egalitarian for all n except $n = 2$.*

To put this result in perspective let us move away for a moment from the category of C^*-algebras. First we can consider real C^*-algebras, which are important in the context of the spectral standard model. Then there are two new cases of non-egalitarian algebras which are $M_2(\mathbb{R})$ and $M_2(\mathbb{H})$. However in the more natural arena of JB-algebras, the exceptional character of these cases disappear, as they all fit into the familly of spin factors. Remember that a spin factor, or Jordan algebra of Clifford type is the direct sum $N \oplus \mathbb{R}$, where N is a real Hilbert space, with Jordan product $(u,x) \circ (v,y) = (xu + yv, xy + \langle u, v\rangle)$. The Jordan algebras of self-adjoint real, complex and quaternionic 2×2 matrices are isomorphic to spin factors with $N = \mathbb{R}^2, \mathbb{R}^3$ and $\mathbb{R}^5$ respectively. It will be proven elsewhere that spin factors are the only simple finite-dimensional non-egalitarian JB-algebras (with the possible exception of the Albert algebra, which is the only case still under investigation).

Returning to the case of C^*-algebras, we now give a classification result.

Theorem 5 *Let I be an isocone in a finite-dimensional C^*-algebra A_F. Then there exists a finite poset P and positive integers $(n_x)_{x\in P}$ such that up to a $*$-isomorphism one has $A = \bigoplus_{x\in P} M_{n_x}(\mathbb{C})$ and $I = \mathrm{Lex}_{x\in P} I_x$.*

Since M_{n_x} is egalitarian if $n_x \neq 2$ and isocones in $M_2(\mathbb{C})$ are easily characterized, as we explained above, this theorem classifies all finite-dimensional I^*-algebras. It shows that the generalization of partial orders given by I^*-algebras is quite limited in the finite-dimensional case: partial order seems to be essentially a commutative phenomenon. However, the new freedom offered by the exceptional $n = 2$ case could prove to be important in the context of the Standard Model, possibly in relation with T-symmetry breaking.

Let us make some comments on the physical interpretation of the above theorem. Recall that we see the elements of I as observables corresponding to generalized clocks. If we admit the usual interpretation rule of quantum physics that eigenvalues of observables corresponds to values that can be possibly measured, then the order on P can be characterized as follows: $x \preceq y$ if and only if no generalized clock ever give a result at x which is strictly larger than its result at y.

On a more mathematical stance, let us note that what the theorem tells us is that the partial order naturally defined on the pure state space by I descends on the structure space, which is P. This was not guaranteed, and is proved by a combinatorial/analytical argument. If there is a conceptual reason for this, we are not aware of it.

We now turn to the important case of almost-commutative algebras, that is, C^*-algebras of the form $\mathcal{A} = \mathcal{C}(M, A_F)$, where A_F is a finite-dimensional algebra. We fix $\mathcal{A}$ an almost-commutative algebra with M compact. For simplicity of exposition, we suppose that $A_F = M_n(\mathbb{C})$ with $n \geq 2$. The case of a general finite-dimensional algebra A_F is not essentially different (we refer to [17] for details). Note that the pure state space of $\mathcal{A}$ is $M \times \mathbb{P}(\mathbb{C}^n)$.

Theorem 6 *Let I be an isocone of $\mathcal{A}$. Then the order induced by I on $M \times \mathbb{P}(\mathbb{C}^n)$ is lexicographic: there exists a partial ordering $\preceq_M$ on M and for each $x \in M$ there exists an isocone I_x of $M_n(\mathbb{C})$ such that :* $(x, \xi) \preceq_I (y, \eta) \Leftrightarrow \begin{cases} x \prec_M y \\ \text{or} \\ x = y \text{ and } \xi \preceq_{I_x} \eta \end{cases}$

The key to prove this theorem is the above classification of finite-dimensional I^*-algebras and the pushforward construction of section 4. Note that the ordering on M is defined from I through the following formula:

$$x \preceq_I y \leftrightarrow \forall a \in I, \max(\sigma(a_x)) \leq \min(\sigma(a_y)) \tag{7}$$

Conversely, given a partial ordering $\preceq_M$ and a map $x \mapsto I_x$ into the isocones of $M_n(\mathbb{C})$, we can state necessary and sufficient conditions for the existence of an isocone of $\mathcal{A}$ inducing them.

Theorem 7 *If M is metrizable and $\preceq$ is a lexicographic order on $M \times \mathbb{P}(\mathbb{C}^n)$ associated to a partial order $\preceq_M$ on M and local isocones $I_x \subset M_n(\mathbb{C})$, then there exists an isocone I in $\mathcal{A}$ inducing $\preceq$ iff*

- *$\prec_M$ is closed.*
- *$L : x \mapsto I_x$ is lower hemi-continuous.*

In that case, I is the lexicographic sum of the isocones I_x over $x \in M$.

This theorem would provide a classification of almost-commutative I^*-algebras (with M metrizable) provided one could prove that the isocone inducing a given order is unique. This would of course follow from the saturation conjecture. Note that it is not difficult to prove that the lexicographic sum $I = \mathrm{Lex}_{x \in M} I_x$ is saturated, that is, it is exactly the set of self-adjoint elements of $\mathcal{A}$ whose Gelfand transform are isotone for $\preceq$. Hence, in the almost-commutative case the saturation conjecture amounts to say that I does not strictly contain any isocone inducing the same ordering on $M \times \mathbb{P}(\mathbb{C}^n)$.

Journal of Physics: Conference Series **634** (2015) 012009 doi:10.1088/1742-6596/634/1/012009

Let us end this section by remarking that the closedness condition on $\prec_M$ is really imposed by the noncommutativity of $M_n(\mathbb{C})$ for $n \geq 2$. If A_F is a more general finite-dimensional C^*-algebra then its structure space will have several sheets. If we consider $\mathbb{C} \oplus M_2(\mathbb{C})$ for instance, there are two sheets M_1 and M_2. Then, the order induced on M_1 which is associated to the commutative algebra $\mathbb{C}$ will not have the closedness restriction.

6. Lorentzian spectral triple and causal cones

Let us now summarize the approach initiated by Franco and Eckstein to the question of causality in noncommutative geometry. We refer to [7] and [19] for details, however we sometimes give indications of proofs which are different than those given there, and might be in some cases easier to generalize.

One first defines Lorentzian spectral "triples", which in fact contains five pieces of information.

Definition 2 *A Lorentzian spectral triple is given by the following data: a pre-C^*-algebra without unit A, a unitization $\tilde{A}$ of A, a Hilbert space H on which A and $\tilde{A}$ are faithfully represented, an unbounded operator D with dense domain such that for all $a \in \tilde{A}$*

- *$[D, a]$ is bounded,*
- *$a(1 + \langle D\rangle^2)^{-1/2}$ is compact, where $\langle D\rangle^2 = \frac{1}{2}(DD^* + D^*D)$.*

and j is a bounded operator such that $j^2 = 1$, $j^* = j$, $[j, \tilde{A}] = 0$, $D^* = -jDj$. Moreover one asks that $j = -[D, T]$ where T is an unbounded operator such that $(1 + T^2)^{-1/2} \in \tilde{A}$.

The role of j is to turn H into a Krein space, and the requirement that j can in fact be written $j = -[D, T]$ is the way Lorentzian signature is single out among arbitrary pseudo-riemannian ones.

The first example is the following: consider a complete and globally hyperbolic Lorentz manifold M, S its spinor bundle, $H = L^2(S)$, $D = -i\gamma^\mu\nabla^S_\mu$ the usual Dirac operator. The algebras A and $\tilde{A}$ are respectively $\mathcal{S}(M)$, the Schwartz functions on M, and the algebra generated by causal bounded functions with bounded derivatives. The role of the operator T is played by a global time function on M, called x^0, so that $j = i\gamma(dx^0)$. On a coordinate chart where x^0 is the first coordinate one can write $j = i\gamma^0$.

Given a lorentzian spectral triple, one can define a causal cone.

Definition 3 *A causal cone $\mathcal{C}$ is a cone in the self-adjoint part of $\tilde{A}$ which contains 1 and such that*

- *$\overline{span_{\mathbb{C}}(\mathcal{C})} = \overline{\tilde{A}}$*
- *$\forall a \in \mathcal{C}$, $j[D, a] \leq 0$ (is a negative semidefinite operator)*

We can then use a causal cone to define a partial order on the space of pure states $P(\tilde{A})$, as one does with an isocone. We can see that we recover the usual partial order on M (which is a subset of the compact space $P(\overline{\tilde{A}})$) through the following result ([7], theorem 11):

Theorem 8 *If $(\mathcal{A}, \tilde{\mathcal{A}}, H, D, j)$ is the Lorentzian spectral triple built out of a globally hyperbolic Lorentzian manifold M, then $f \in \tilde{\mathcal{A}}$ is causal if and only if $j[D, f] \leq 0$.*

This can be most easily proved using a pseudo-orthonormal moving frame (vierbein). We can then define the Lorentz gradient of f by $df(v) = g(v, \nabla f)$, and decompose ∇f in the form $-(\partial^0 f)e_0 + (\vec{\nabla} f)^i e_i$ in the moving frame. We have $i\gamma^0[D, f] = -\partial^0 f\mathrm{Id} + b$, with $b = \sum_{1\leq i\leq 3} \partial^i f\gamma^0\gamma^i$.

Noting that $b^* = b$ and $b^2 = \|\vec{\nabla} f\|^2\mathrm{Id}$, we see that the spectrum of b is $\{\pm\|\vec{\nabla} f\|\}$. Hence $i\gamma^0[D,f]$ is negative semidefinite iff ∇f_x lies inside the past-cone at every $x \in M$, which is easily seen to be a necessary and sufficient condition for f to be causal.

Now since globally hyperbolic manifolds are toposets, we see that the order defined by

$$x \leq_M y \Leftrightarrow (\forall f \in \tilde{\mathcal{A}}, j[D,f] \leq 0 \Rightarrow f(x) \leq f(y)) \tag{8}$$

is exactly the original causal order of the manifold.

7. Comparison of the two approaches

We see that the two approaches agree as far as (globally hyperbolic) Lorentzian manifolds are concerned: a causal cone is just the smooth part of the isocone of causal functions. To see what happens in the noncommutative case, we will focus on a simple though enlightening example. We take $M = \mathbb{R}^{1,1}$ to be the 2-dimensional Minkowski spacetime. We consider the finite-dimensional spectral triple (A_F, H_F, D_F) with $A_F = M_2(\mathbb{C})$, $H_F = \mathbb{C}^2$, $D_F = \mathrm{diag}(d_1, d_2)$. A product lorentzian spectral triple is defined by $\mathcal{A} = \mathcal{S}(M)\otimes A_F$, $\mathcal{H} = L^2(S)\otimes H_F$, where S is the spinor bundle which here is trivial with fiber $\mathbb{C}^2$, $D = D_M \otimes 1 + \gamma^0\gamma^1 \otimes D_F$, with $D_M = -i\gamma^\mu\partial_\mu$, and $j = i\gamma^0 \otimes 1$. The algebra $\tilde{\mathcal{A}}_M$ is defined to be the linear span of smooth causal bounded functions with bounded derivatives, and $\tilde{\mathcal{A}} = \tilde{\mathcal{A}}_M \otimes A_F$. This in fact corresponds to the Penrose compactification of the Minkowski plane which we briefly recall. Let (x^0, x^1) be the canonical coordinates on $\mathbb{R}^{1,1}$, and write $u = x^0 + x^1$, $v = x^0 - x^1$. The map $\phi : M \to P =]-\pi;\pi[^2$ defined by $\phi(u,v) = (\mu,\nu) := (2\tan^{-1}(u), 2\tan^{-1}(v))$ is a smooth diffeomorphism (the factor of 2 is inserted to stick with usual conventions). Moreover ϕ sends the causal order $\leq_M$ on M, which in the coordinates (u,v) is just the product order on $\mathbb{R}^2$, to the product order $\leq_P$ on P. Since the derivatives of ϕ are all bounded, the push-forward $\phi_* : f \mapsto f\circ\phi^{-1}$ sends $\tilde{\mathcal{A}}_M$ to the linear span of bounded causal (for $\leq_P$) functions on P with bounded derivatives. Such functions can be extended by continuity on $\bar{P} = [-\pi;\pi]^2$. To see it notice that causal functions on P can be expressed in the form $g(\mu) + h(\nu)$ with g, h non-decreasing on $\mathbb{R}$. Using the Stone-Weierstrass theorem we then immediately see that the linear span of bounded causal functions on P with bounded derivatives are dense in $\mathcal{C}(\bar{P})$. Hence the pure state space of $\tilde{\mathcal{A}}_M$ is homeomorphic to the Penrose compactification of M.

Let us look for the elements of the causal cone. A hermitian-matrix-valued function α is in $\mathcal{C}$ if $j[D,\alpha] = \gamma_0\gamma_\mu \otimes \partial^\mu\alpha - \gamma_1 \otimes i[D_F,\alpha] \leq 0$. With $\gamma_0 = \begin{pmatrix} 0 & i \\ i & 0 \end{pmatrix}$, $\gamma_1 = \begin{pmatrix} 0 & -i \\ i & 0 \end{pmatrix}$ this condition can be written in block-matrix form:

$$\begin{pmatrix} 2\partial_u\alpha & [D_F,\alpha] \\ -[D_F,\alpha] & 2\partial_v\alpha \end{pmatrix} \geq 0 \tag{9}$$

In particular this condition implies that

$$\partial_u\alpha \geq 0, \partial_v\alpha \geq 0 \tag{10}$$

Integrating (10) we see that if $x \leq y$ (where $\leq$ denotes the usual causal order on Minkowski space), then $\alpha(x) \leq \alpha(y)$ as operators. Conversely, if $\alpha(x) \leq \alpha(y)$ for all $\alpha \in \mathcal{C}$, then using $\alpha = \mathrm{diag}(t_1, t_2)$, where t_1, t_2 are two time functions, one sees that $x \leq y$. Hence we have proved that

$$x \leq_M y \Leftrightarrow \forall\alpha \in \mathcal{C}, \alpha(x) \leq \alpha(y) \tag{11}$$

This formula can be put in contrast with (7) (where, we recall, $\preceq_M$ cannot be equal to $\leq_M$). In particular we note that if $\lambda_1(x) \leq \lambda_2(x)$ are the eigenvalues of $\alpha(x)$, then (10) shows that

$x \mapsto \lambda_i(x)$ are causal functions. It is however perfectly possible that $\lambda_2(x) > \lambda_1(y)$ with $x \leq_M y$. If we keep the eigenvalue/observed value correspondance and interpret the elements of $\mathcal{C}$ as causal observables, we have here a generalized clock which can assign a later date to x than to y, which is inconsistent with the observations of section 5. This shows that a causal cone is not an isocone in the noncommutative case. The mathematical origin of the difference between causal cones and isocones is that the former are not stable under non-decreasing functional calculus (this can be traced back to the fact that monotone functions are not necessarily operator-monotone).

Finally we recall the main result of [19], which is the characterization of the order induced on the pure states of the form $(x, \omega_{[\xi]})$, with $x \in M$ and $\omega_{[\xi]}$ is the vector state $a \mapsto \langle \xi, a\xi \rangle$ on $M_2(\mathbb{C})$. Let us fix some notations before stating the theorem: $d_{NC}(p,q)$ is Connes' noncommutative distance between two pure states on $M_2(\mathbb{C})$ induced by D_F, and $\ell(x,y)$ is the Lorentz distance between x and y, that is to say the supremum of $\int_\gamma ds$ where γ runs over all future-directed causal curves from x to y (it is set to 0 if y is not in the future of x). In the case at hand, $\ell(x,y)$ is the proper time elapsed from x to y measured by an inertial observer.

Theorem 9 $(x, \omega_{[\xi]}) \preceq_C (y, \omega_{[\eta]})$ *iff* $x \leq_M y$ *and* $\ell(x,y) \geq d_{NC}(\omega_{[\xi]}, \omega_{[\eta]})$.

Note that the inequality implies that $d_{NC}(\omega_{[\xi]}, \omega_{[\eta]})$ is finite. We recall that it is the case iff $\omega_{[\xi]}$ and $\omega_{[\eta]}$ are at the same latitude on the 2-sphere, where the north and south poles are given by the eigenstates of D_F (see [20]). Let us explain briefly why the condition is necessary. First we can suppose without loss of generality that $x = (0,0)$ and $y = (s,0)$. Then $\ell(x,y) = s$. Using matrix-valued functions of the form $\alpha(t,r) = tI_2 + A$, where A is a constant hermitian matrix, we easily obtain from (9) that $\alpha \in \mathcal{C}$ iff $\|[D_F, A]\| \leq 1$, which gives the result.

We now turn our attention to I^*-algebras. We work directly with the Penrose compactification, so the algebra is $\mathcal{A} = \mathcal{C}(\bar{P}) \otimes M_2(\mathbb{C})$. The other datum we need is an isocone $\mathcal{I}$ of $\mathcal{A}$. Note the difference with the previous approach: the usual causal order was encapsulated in the choice of D and j, wheras here it is induced by the isocone, and as such cannot be the usual causal order on $\bar{P}$, since the noncommutative nature of the algebra $\mathcal{A}$ imposes that the causal relations locally vanish. What we ask then, is that the isocone-induced order $\preceq_\Lambda$ on $\bar{P}$ depends on a constant Λ and tends to the usual causal order when $\Lambda \to 0$. If we moreover ask $\preceq_\Lambda$ to be Lorentz-covariant, it will be easily constructed.

We first define $\preceq_\Lambda$ on $\mathbb{R}^{1,1}$ and then we will send it to $\bar{P}$. On $\mathbb{R}^{1,1}$ we set:

$$(x^0, x^1) \preceq_\Lambda (x'^0, x'^1) \Leftrightarrow x^0 \leq x'^0 \text{ and } (x^0 - x'^0)^2 - (x^1 - x'^1)^2 \geq \Lambda^2 \qquad (12)$$

That is, we simply replace forward light-cones with forward hyperbolae of mass Λ. Changing to the coordinates (u,v), and using the notations $\Delta u = u - u'$, $\Delta v = v - v'$, we easily see that this definition is equivalent to

$$(u,v) \preceq_\Lambda (u',v') \Leftrightarrow \Delta u \geq 0, \Delta u \Delta v \geq \Lambda^2 \qquad (13)$$

Hence in P, using the same notation to denotes the partial order we obtain

$$(\mu,\nu) \preceq_\Lambda (\mu',\nu') \Leftrightarrow \Delta\mu \geq 0, (\Delta \tan\frac{\mu}{2})(\Delta \tan\frac{\nu}{2}) \geq \Lambda^2 \qquad (14)$$

with obvious notations, which is extended to $\bar{P}$ by $(\mu,\nu) \preceq_\Lambda (\mu',\nu') \Leftrightarrow \Delta\mu \geq 0, \Delta\nu \geq 0$ anytime (μ,ν) or (μ',ν') belongs to the boundary.

To complete the data we have to choose a lower-hemicontinuous isocone map $(\mu,\nu) \mapsto I(\mu,\nu) \subset M_2(\mathbb{C})$. For instance we can take a constant map with value a non-trivial isocone in $M_2(\mathbb{C})$ defined by a closed and geodesically convex subset K of the sphere S^2. We finally define $\mathcal{I} = \mathrm{Lex}_{(\mu,\nu)\in\bar{P}} I(\mu,\nu)$, with $I_x = I$ =constant, and $\bar{P}$ ordered by $\preceq_\Lambda$. We illustrate the two situations in figure 1.

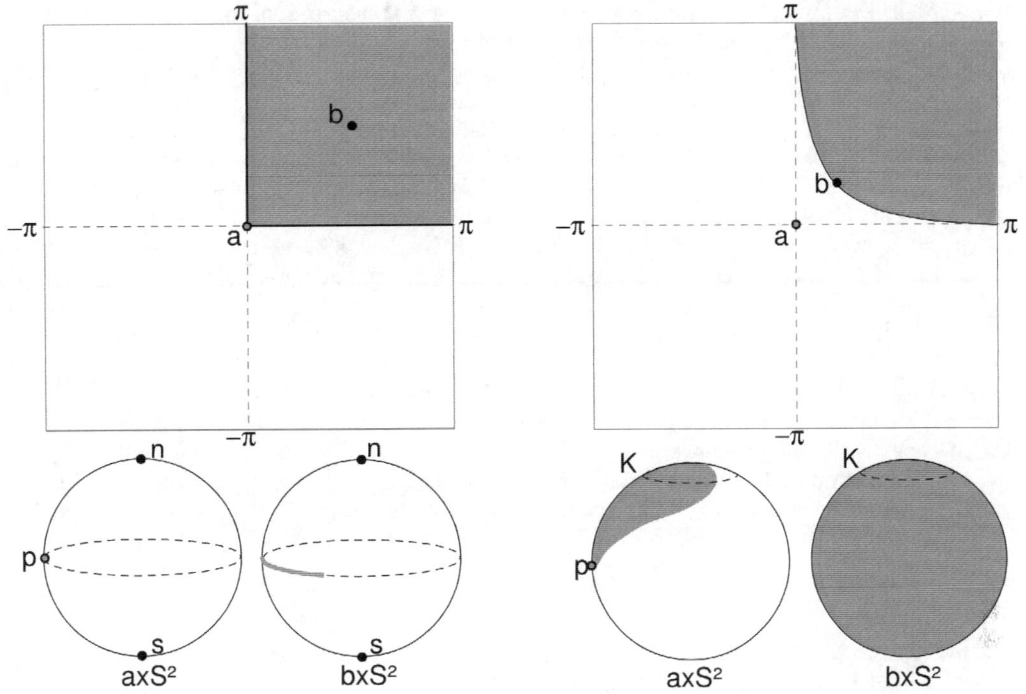

Figure 1. This figure shows the set of pure states which are larger than (a, p), where a is an event in the Penrose compactification of the Minkowski plane, and p is a pure state in $P(M_2(\mathbb{C})) \simeq S^2$, for the two order relations described in the text: on the left side the order induced by a causal cone, and on the right from an isocone. Each point in the Penrose square represents a 2-sphere. It is depicted in grey if the sphere contains at least a pure state larger than (a, p). Under the square the two spheres at a and b are shown. On the left, only the states lying at the same latitude than p can be in the future of (a, p). The North and South poles are given by the eigenprojectors of the Dirac operator. In the sphere $b \times S^2$ one sees that the states in grey are those whose geodesic distance from p is less that the Lorentz distance from a to b. On the right we can see a non-trivial "inner" future light-cone in the sphere $a \times S^2$. It is easily seen to be the intersection of S^2 with $p + K^\circ$, where K° is the dual cone of K, which is here taken to be a spherical cap. At b or any other point in the future of a and not equal to a, the sphere is entirely in the future of (a, p).

8. Conclusion

It is evident from the above examples that there is much more freedom in the isocone approach, and it is not surprising since isocones generalize partial orders which can be arbitrary apart from possesing the toposet property. In particular we do not have any axiom which would ensure the compatibility with a given metric, incarnated in the noncommutative setting by the Dirac operator. This is clearly an important ingredient to add in the future development of the theory. On the other hand we believe that the approach is sufficiently general and physically well motivated to be trusted on its qualitative conclusions, in particular on the necessary vanishing of the causal relations at small scale (with no indication on the smallness scale, though). As a matter of fact, equivalent conclusions have been reached in [21] and [22] by studying the properties of the spectral action at high energy. The agreement between the isocone approach and the properties of the spectral action is all the more striking that the latter is formulated in the euclidean formalism. However, this action reproduces, under some cut-off energy scale, the one of the Standard Model which we know is correct on experimental ground. An optimistic explanation of the agreement between the disappearance of causality at small scale/high energy, as required by the isocone approach, and the non-propagation of high-energy bosons implied by the spectral action formalism could therefore be the following one: 1) the fundamental hypotheses laid up in the first section can be trusted (i.e. Nature really behaves like this), 2) the asymptotic developments of the spectral action can be trusted, 3) Nature is consistent with itself. To temperate this enthousiasm, let us observe that the local disappearance of causality implied by the isocone approach might be more soberly viewed as just another theorem of the noncommutativity-implies-discreteness kind that we are familiar with since the advent of quantum theory.

However, as we have seen, causal cone theory does not predict such a discreteness. Quite the contrary, it is fully compatible with usual causality at all scale. Does it mean that it is wrong-headed ? Not necessarily. In fact it is interesting to note that the bottom-up isocone approach and the top-bottom causal cone theory each has what the other lacks. The first has, we believe, a clear physical interpretation and is able to derive sensible qualitative conclusions at large energies, the second has a direct contact with the usual apparatus of noncommutative geometry, providing in particular a very transparent connection between the Dirac operator and the causal order. We think that it is not only possible, but necessary, that the two approaches merge. It is all too obvious that the culprit of their present incompatibility is Minkowski spacetime, whose smoothness cannot be trusted at arbitrary large energy scale. In replacing smooth spacetime with a discrete structure we do not expect to encounter any difficulty on the isocone side, but we will have to discretize the Dirac operator along the lines of [23]. Whether this permits to equate causal cones with isocones is under investigations.

References

[1] Chamseddine AH, Connes A, Marcolli M 2007, Gravity and the Standard Model with neutrino mixing, *Adv. Theor. Math. Phys.* **11** 991

[2] van den Dungen K, Paschke M, Rennie A 2013, Pseudo-Riemannian spectral triples and the harmonic oscillator, *J. Geom. Phys.* **73** 37

[3] Besnard F 2007, *Canonical quantization and the spectral action, a nice example*, *J. Geom. Phys.* **57** 1757

[4] Doplicher S, Fredenhagen K, Roberts JE 1995, The quantum structure of spacetime at the Planck scale and quantum fields, *Commun. Math. Phys.*, **172** 187

[5] Bombelli L, Lee J, Meyer D, Sorkin RD 1987, Space-time as a causal set, *Phys. Rev. Lett.* **59**, 521

[6] Besnard F 2009, A noncommutative view on topology and order, *J. Geom. Phys.* **59** 861

[7] Franco N, Eckstein M 2013, An algebraic formulation of causality for noncommutative geometry, *Class. Quant. Grav.* **30** 135007

[8] Besnard F 2011, Time of philosophers, time of physicists, time of mathematicians, *Preprint* arXiv:1104.4551

[9] Muga JG, Sala Mayato R, Egusquiza IL (Eds.) 2002, *Time in quantum mechanics*, (Berlin: Springer)

[10] Minguzzi E, Sànchez M 2008, The causal hierarchy of spacetimes, *Recent developments in pseudo-Riemannian geometry*, ESI Lect. Math. Phys 299
[11] Emch GG 1984, *Mathematical and Conceptual Foundations of 20-th Century Physics*, (Amsterdam: Elsevier)
[12] S. T. Ali, M. Engliš, *Quantization methods: a guide for physicists and analysts*, Rev. Math. Phys. **17** (2005) 391 math-ph/0405065
[13] Rovelli C 2002, GPS observables in General Relativity, *Phys. Rev. D* **65** 044017
[14] Sheppard B 2001, A Stone-Weierstrass theorem for postliminal JB-algebras. *Q. J. Math.* **52** 507
[15] Besnard F 2013, Noncommutative ordered space: examples and counterexamples, *to appear in Class. Quantum Gravity*, arXiv:1312.2442
[16] Nachbin L 1965, *Topology and Order*, (Princeton: D. Van Nostrand Company)
[17] Bizi N, Besnard F 2014, The disappearance of causality at small scale in almost-commutative manifolds, *Preprint* arXiv:1411.0878
[18] Kadison R, Ringrose J 1983, *Fundamentals of the Theory of Operator Algebras*, (New York: Academic Press)
[19] Franco N, Eckstein M 2014, Exploring the Causal Structures of Almost Commutative Geometries, *SIGMA* **10** 010
[20] Iochum B, Krajewski T, Martinetti P 2001, Distances in Finite Spaces from Noncommutative Geometry, *J. Geom. Phys.* **37**, 100
[21] Kurkov MA, Lizzi F, Vassilevich D 2014, High energy bosons do not propagate, *Phys. Lett. B* **731**, 311
[22] Alkofer N, Saueressig F, Zanusso O 2014, Spectral dimensions from the spectral action, *Preprint* arXiv: 1410.7999
[23] Marcolli M, van Suijlekom WD 2014, Gauge networks in noncommutative geometry, *J. Geom. Phys.* **75**, 71

Institute of Physics
Dirac House, Temple Back
Bristol BS1 6BE UK

CURRAN ASSOCIATES INC
proceedings
.com